George Pitt-Rivers and the Nazis

George Pitt-Rivers and the Nazis

BRADLEY W. HART

Bloomsbury Academic
An imprint of Bloomsbury Publishing Plc

B L O O M S B U R Y
LONDON · NEW DELHI · NEW YORK · SYDNEY

Bloomsbury Academic

An imprint of Bloomsbury Publishing Plc

50 Bedford Square	1385 Broadway
London	New York
WC1B 3DP	NY 10018
UK	USA

www.bloomsbury.com

BLOOMSBURY and the Diana logo are trademarks of Bloomsbury Publishing Plc

First published 2015

British Library Cataloguing-in-Publication Data
A catalogue record for this book is available from the British Library.

ISBN: HB: 978-1-4725-6995-0
PB: 978-1-4725-6994-3
ePDF: 978-1-4725-6996-7
ePub: 978-1-4725-6997-4

Library of Congress Cataloging-in-Publication Data
Hart, Bradley W.
George Pitt-Rivers and the Nazis/Bradley W. Hart.
pages cm
Includes bibliographical references and index.
ISBN 978-1-4725-6995-0 (hardback) – ISBN 978-1-4725-6994-3 (paperback) –
ISBN 978-1-4725-6996-7 (ePDF) – ISBN 978-1-4725-6997-4 (ePub) 1. Pitt-Rivers, George
Henry Lane Fox, 1890–1966. 2. Pitt-Rivers, George Henry Lane Fox, 1890-1966–
Political and social views. 3. Anthropologists–Great Britain–Biography. 4. Gentry–Great
Britain–Biography. 5. Politicians–Great Britain–Biography. 6. Eugenics–Great Britain–
History–20th century. 7. Fascism–Great Britain–History–20th century. 8. Antisemitism–
Great Britain–History–20th century. 9. Churchill, Clementine, 1885-1977–Family.
10. Great Britain–Politics and government–1936-1945. I. Title.
DA566.9.P55H37 2015
942.3'32082092–dc23
[B]
2014047442

Typeset by Integra Software Services Pvt. Ltd.

For Val and Anthony

CONTENTS

LIST OF ILLUSTRATIONS

ACKNOWLEDGEMENTS

Any project acquires a number of debts along the way, and the present work has incurred more than most. First, and most importantly, this book would have been impossible without the incredible generosity of Anthony and Valerie Pitt-Rivers, to whom it is gratefully dedicated. From my first contact with them in 2009, they have shown me incredible kindness and in 2010 made the decision to donate the George Pitt-Rivers archive to the Churchill Archives Centre, Cambridge. These materials have provided the archival base for this book and will undoubtedly feature in the work of other scholars over the years and decades to come. In addition, Mr Pitt-Rivers made an extraordinary effort to help me during my research and writing, saving me from many mistakes along the way and giving me context beyond the archival materials themselves.

In a similar vein, others have shown immense kindness along the way as well. Michael Taylor kindly allowed me access to materials related to his mother's life and career that were hugely valuable. The discovery of the South African aspect of this project relatively late in the research process has added an unexpected and entirely new dimension of interest, and I am extremely grateful for the opportunity to tell this important part of the story.

This project also relied on access to a number of archives and collections in the United Kingdom, the United States, Germany and elsewhere. Allen Packwood and his outstanding staff at the Churchill Archives Centre not only took in the Pitt-Rivers collection when it was offered but also did an outstanding job with its preservation. Thanks to their efforts, these materials will be available to researchers in perpetuity. In addition, Churchill College once again became my home for much of this research, and I am grateful to the former master, Sir David Wallace, and the fellows of the college for granting me a visiting by-fellowship during Easter Term 2014. Without support from the Archives Centre and the College, this project would never have been completed.

There are many others to thank as well. My doctoral supervisor, Sir Richard J. Evans, made many helpful suggestions over the years that gave me both wider perspectives on my work and taught me how to write history. Richard Carr read early drafts of this work and provided many valuable suggestions along the way, all of which gave me valuable perspective and direction. My collaborative work with Landis MacKellar gave me the opportunity to explore wider aspects of Pitt-Rivers' career and his encouragement led me to pursue a number of loose ends that proved significant. Dan Stone provided numerous suggestions throughout the review process that were incredibly valuable. My colleagues at California State University, Fresno provided important support throughout the research and writing process, for which I am grateful. At Bloomsbury, Frances Arnold and Emma Goode provided critical support for the project and made a great effort to bring it to a successful conclusion.

Finally, I would be remiss to not mention the important contribution made by my family to this work. My parents, Dennis and Sharon, provided support in so many ways for this project and my past undertakings. Likewise, my grandparents have always been among my biggest supporters, and I owe them immensely.

In closing, I should note that all editorial decisions related to this work have been made by me alone. I have not been asked to omit any details or information by any third party, and any decisions to do so have been made by me. Similarly, while I have asked for some factual clarifications from the family members of individuals discussed in this work, I alone have made the final judgement on what information to include under no external influence. Accordingly, any remaining errors of fact are mine alone. At several points, I have included only details for which there is extant documentary proof and have omitted aspects of the narrative of which I have only hints but have been unable to find primary source corroboration. Future scholars may well refine and build this narrative as additional sources become available.

Bradley W. Hart

The Pitt-Rivers and Stanley families
*Showing only selected members (others omitted for clarity)

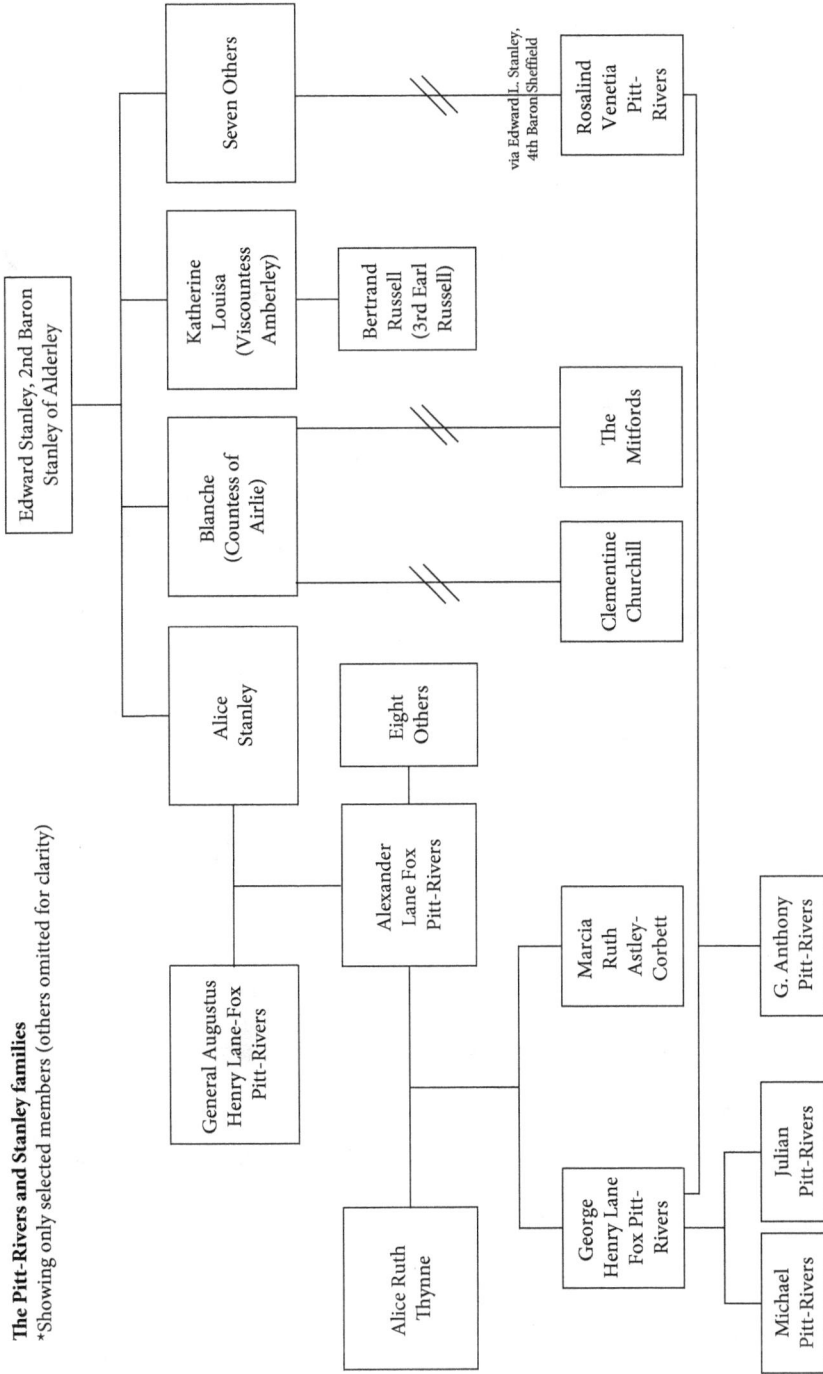

Edward Stanley, 2nd Baron Stanley of Alderley

Alice Stanley

Blanche (Countess of Airlie)

Katherine Louisa (Viscountess Amberley)

Seven Others

General Augustus Henry Lane-Fox Pitt-Rivers

Alexander Lane Fox Pitt-Rivers

Eight Others

Clementine Churchill

The Mitfords

Bertrand Russell (3rd Earl Russell)

via Edward L. Stanley, 4th Baron Sheffield

Rosalind Venetia Pitt-Rivers

Alice Ruth Thynne

George Henry Lane Fox Pitt-Rivers

Marcia Ruth Astley-Corbett

G. Anthony Pitt-Rivers

Michael Pitt-Rivers

Julian Pitt-Rivers

Introduction

The morning of 27 June 1940 began unremarkably in the rural Dorset village of Hinton St Mary. There was, of course, a new normalcy to things: Britain was once again at war, and just five days before the French had surrendered to Adolf Hitler's seemingly unstoppable armies. It was now only a matter of time before the planes and bombers that had helped pound the French army into submission and levelled the centre of Rotterdam would be used against Britain and, ultimately, London. Depending on how the situation developed over the coming weeks and months, it might only be a matter of time before the German army arrived on British soil. A little more than a week before, Prime Minister Winston Churchill had told the House of Commons and the country that 'the battle of Britain is about to begin'.[1]

In Hinton's grand manor house, the squire was sitting down to lunch. A tall, slight man in his early fifties sporting a cavalry officer's moustache, he walked with a limp and carried a heavy wooden cane. His leg and hip were in pain, as they were every day, and the feeling was a daily reminder of the wound he had suffered in the last war. It had been called the 'war to end all wars' and yet now there was another world war maiming and killing a new generation of Britain's young men. Two of his three sons were already in uniform, just as he had been in 1914.

At the luncheon table, he was soon joined by several officers from the army unit billeted in his house and in the nearby village. The officers were polite but reserved: conversations with their host were not always enjoyable, and there were sometimes things said that made them uncomfortable. At least one of them had already told the Dorset chief constable about some of the remarks he had heard in the house. Also joining them that day was a well-known former Navy man: Admiral Sir Barry Domvile, former commander of the HMS *Royal Sovereign*, Director of Plans, and, later, Director of the Department of Naval Intelligence. He and his German-born wife had been staying in the house for the past few days.

The normalcy of the day did not last long. One guest did not arrive for the meal, instead receiving orders to set up his Bren machine gun in the garden with the muzzle pointing menacingly towards the front door. Soon there were cars seen approaching the house. In one were investigators from the Dorset Constabulary, accompanied by uniformed and plain-clothes police who would shortly search the house from top to bottom. Placing the manor's owner under arrest, the constable presented him with the reasons for his detention: 'The Secretary of State has reasonable cause to believe that you have been a member of the organization now known as British Union; or have been active in the furtherance of its objects; and that it is necessary to exercise control over you.'[2] The squire could hardly find the words to answer. 'Is this the reward for patriotism?' he muttered. Domvile described him as 'trying to be brave, but looking ghastly – and babbling'.[3]

There was little time to talk. The police search of the house began immediately. It was only in the study and office that anything of interest was found. In addition to a large quantity of correspondence, they recovered a plaque of Adolf Hitler, photographs of other Nazi leaders and four small pennants 'apparently for use on a motor car'. They were the national flags of Italy, Spain and Nazi Germany – plus the Union Jack.[4] The evidence thus secured, the manor's owner was driven to London. He would not again spend significant time in his home until the end of the war in 1945. Domvile and his wife remained at the house and would shortly return home, only to be arrested themselves days later. Among the things left behind after the police raid was an 'elegantly bound presentation copy' of Hitler's autobiographical tract *Mein Kampf* in the library. Throughout the war, the book 'was occasionally taken out and looked at, almost reverentially; then put back, without ever being damaged or mutilated' by the soldiers stationed there.[5]

From Dorset, the owner of the manor was taken to Brixton Prison and later to Ascot, where he was detained in a camp for much of the autumn. Suffering pain from his war wounds under the harsh conditions, he was returned to Brixton to pursue appeals against his continued detention. There would be no charges filed and no trials held, for under Britain's wartime regulations, it was perfectly possible to hold someone indefinitely on the mere suspicion that they might pose a threat to national security presently or in the future. It would not be until early 1942 that he would actually be allowed to leave the confines of physical detention, and even then the authorities would keep a close eye on him.

The foregoing account is, to a certain extent, how this story ends rather than how it begins. The arrest of George Henry Lane Fox Pitt-Rivers ('Jo', to those who knew him best) on 27 June 1940 took place at a moment of great peril for Britain: with the fall of France, German bombing of the country was only weeks from beginning and ground invasion might soon follow. Pitt-Rivers had established a reputation for his pro-Hitler views by the start of the war, writing extensively on international affairs throughout the 1930s, publicly endorsing Hitler's claims to the Sudetenland that ended in the Munich Agreement, and praising Francisco Franco for his fight against communism in Spain. When it came to his arrest and detention, however, there was never a question about him posing an actual threat to national security. Both MI5 and the Home Office were aware that there was no evidence he had ever actively helped the Germans or knowingly given sensitive information. The judgement of historian A.W. Brian Simpson was probably in line with what many contemporaries thought of Pitt-Rivers: he was 'an appalling bore, with academic pretensions' who was 'somewhere between eccentric and dotty' but nonetheless 'harmless'.[6]

This uncharitable description might well have been true by the time of Pitt-Rivers' arrest in 1940, but it was certainly not the case a decade earlier. In the late 1920s, the 'academic pretensions' mentioned would have been well deserved through both his work and the circle of friends he had cultivated. While Pitt-Rivers would never have achieved the accomplishments and fame of his colleagues who became the legends of the twentieth-century anthropology – Bronislaw Malinowski, Raymond Firth, C.G. Seligman, E.E. Evans-Pritchard and others – he possessed the analytic mind for scholarship and had an iconoclastic streak that Malinowski, among others, admired and believed was one of his greatest academic assets. If George Pitt-Rivers' life had been cut short in 1929, he would have been mourned at Britain's leading universities as a tragic figure whose potential had been unfairly curtailed by the vagaries of fate.

The question of how and why this promising academic career was so abruptly derailed and ultimately destroyed is at the heart of this book. It is my fundamental contention that George Pitt-Rivers had the unrealized potential to be remembered as one of the greatest anthropologists of the twentieth century, much as his grandfather, General Augustus Lane-Fox Pitt-Rivers, is now seen with similar adulation in the century prior. The general has often been referred to as the father of British archaeology, and while this description is an exaggeration, there is no doubt that his work was renowned and impactful.[7] The Pitt-Rivers Museum in Oxford stands as a lasting monument to his legacy, and the shadow of the general's work hung over his grandson throughout his life. The book that George Pitt-Rivers should always have written was a biography of his famous grandfather, and he jealously guarded the personal papers and journals that would have made such a project possible. As will be seen, it was ultimately a combination of his ego and his deep pessimism that prevented him from doing so.

Writing a biography always carries unique pitfalls: the temptation can be to alternatively lionize the subject and exaggerate their importance or, in contrast, to apply hindsight and exclusively illustrate their retrospective folly. In some ways, this project has carried with it additional dangers of its own due to the nature of the subject. There can be no doubt cast on some facts about George Pitt-Rivers' life. Neither of his marriages were successful and, like his grandfather, he had a poor relationship with most of his family. He was often irascible and personally difficult, as his correspondence make clear. He held and expressed political and racial views that are rightly seen today as reprehensible. He could be extremely pretentious despite his own modest accomplishments in most areas, and he considered himself, at various points, to be a philosopher, art critic, scientist and politician, though he would always zealously deny having any desire to be the last of these.

On the other hand, Pitt-Rivers was undoubtedly intelligent and no one, including his enemies, ever denied his aptitude and scholarly abilities. He was erudite and charming, particularly with women, and his upper-class sensibilities and manners impressed many he encountered. He had a high opinion of himself and his work but was rarely as pompous as he might have been, and he was often described by those who knew him best with adjectives that included 'kind', 'generous' and 'understanding'.[8] There were reasons for his successes, however short-lived they might have been. The aged anthropologist Arthur Keith was remarkably astute when he described the young Pitt-Rivers he met in 1926 as 'clearly a young autocrat – with nothing of the democrat in his composition'.[9]

Despite this potential, Pitt-Rivers would never rise to the academic heights that he might have been able to achieve. In large part, this was because despite his intelligence, he harboured an obsession that would consume him in the 1930s: anti-Semitism. Nowhere in his papers or correspondence is an explanation offered for where and when his obsession with Jews began; it appears to have simply been a first principle and underlying premise for his arguments and beliefs. This is somewhat unusual: as an introspective academic, Pitt-Rivers recorded and discussed his intellectual development often. He documented, for instance, the blossoming of his interest in the ideas of Friedrich Nietzsche, Sigmund Freud, Thomas Malthus, Benito Mussolini and even Adolf Hitler. Correspondence survive in which he described first encountering the ideas of eugenics as a young man at Eton and having his conscience stirred by them, and in long letters he laid out his reasons for believing in the legalization of birth control and making it accessible to all couples. His life-long and often obsessive opposition to communism is even documented from a young age onwards. It is almost always possible to trace the development of his thought and examine how he contended with events that challenged his ideas.

This is not the case with his anti-Semitism. There is no document spelling out the reasons or philosophical justifications for his negative and conspiratorial views towards the Jews. He evidently saw no need to justify or explain his analyses in this area. It is certainly true that anti-Semitism was widespread among men and women of his generation, particularly in aristocratic circles. Kazou Ishiguro's 1989 novel *The Remains of the Day* and its subsequent film adaptation have illustrated the late-1930s world of appeasers, covert Nazi sympathizers and both casual and endemic anti-Semitism for modern audiences, and in some senses Pitt-Rivers could almost be a character from its pages.[10] Yet, for a man who saw himself first and foremost as a scientific practitioner, he appears to have engaged in little critical examination of his own views towards the Jews.

In addition to the virulent anti-Semitism that he was never able to move beyond, there were other factors for Pitt-Rivers' failure in the academic realm. When his father died in 1927, he had no realistic option but to assume responsibility for his family's vast estate. He had a sister but no brothers to help with estate management duties, and his sons were not yet old enough to take on a major role. Duty dictated that the estate and the traditional role of the squire become his primary responsibilities and academia be relegated to a secondary concern. While he was initially able to use his new assets to great effect by hosting major scientific gatherings at his estate, these events were eventually replaced by far-right political meetings. The precarious state of British agriculture soon led Pitt-Rivers to embark on his first and only foray into politics in the 1935 General Election. He not only failed to win the parliamentary seat he desired, but he also disastrously courted and received the support of Oswald Mosley's British Union of Fascists (BUF) for his campaign. It was one thing to have an academic interest in fascism and extol its merits in a lecture room or in the confines of a Piccadilly club, as he was known to do, but it was very different to coordinate with Mosley directly. The Security Service (MI5) was increasingly taking note of his activities, and the exact details of what might or might not have taken place in his secret talks with the BUF were given great significance in the hearings that took place after his arrest.

Pitt-Rivers therefore never came close to reaching the extent of his academic potential. The insidious nature of his anti-Semitic beliefs, coupled with the diversions of first estate management and, later, the allure of 'peacemaking' to avoid another war with Germany, effectively ruined his reputation by drawing him into the circles of the Nazi-sympathizing far right. During his wartime internment, he arrogantly wrote to Worcester College, Oxford, asking for academic sanctuary there to continue his work. This idea was rejected, and his academic career was essentially over. His final published work, which appeared posthumously, was a poorly reviewed analysis of Christian iconography that sought to disprove the idea that the Hinton St Mary mosaic – discovered on land he had previously owned in 1963 – contained an image of Jesus.[11] He was iconoclastic to the end, never accepting the judgements of those with whom he disagreed regardless of their station or learning.

There were many rumours about Pitt-Rivers in his own time and after. In the Dorset villages near his estate, it was scurrilously rumoured that he had endowed a school for the illegitimate children he fathered. Some villagers alleged that before the war, he laid out crops or hedges in giant arrows so that when the plants were in bloom and viewed from above they would point towards London and guide lost German bombers to their targets. A former household employee claimed that in the house, there was a secret hole with a radio transmitter inside that he used to contact Germany. Swastikas were carved into his desk and fireplaces and woven into rugs, the employee claimed.[12] MI5 itself reported that he had been seen wearing a gold swastika badge at a club in Central London and was seen leaving pro-Nazi propaganda between the pages of books in the library there. After the war began, the chief constable of Dorset claimed he had a habit of walking around his estate wearing a Nazi Party armband and talking about how quickly he hoped the Germans would reach Paris.[13]

Most, if not all, of these rumours were false. There is no evidence to suggest that Pitt-Rivers ever fathered an illegitimate child, and searches of his home after his arrest in 1940 turned up no evidence of a swastika motif in the decor. If it was the Athenaeum Club where Pitt-Rivers had supposedly distributed fascist literature – he joined in the mid-1930s – there appears to be no surviving record of anyone ever reporting the matter or asking him to desist. On the other hand, as is often the case, there was a certain kernel of truth in each, however small. There is no doubt that Pitt-Rivers had a lifelong interest in women and had a number of female companions. As already noted, investigators searching his home in 1940 did in fact find a plaque of Hitler and a small German flag, but they certainly would have remarked on the addition of swastikas to the architecture or décor of the house.[14] On the matter of the arrows being planted in the hedgerows or crops, even many of the locals were sceptical, with one rejecting the idea on the grounds that '"he couldn't have planted a crop" to save his life'.[15]

This work seeks to go beyond these rumours and speculations to present an account of George Pitt-Rivers' life and career rooted in the existing evidence. Throughout it, I make no effort to 'rehabilitate' him or his views. If editorial commentary seems to be lacking in parts it is merely because his documented statements speak for themselves and do not require amplification. My wider intention in this book is to present an example of the political and social roads that were possible to travel in the first half of the twentieth century and the reasons why some choose to take the paths that they did. Pitt-Rivers did not fall into the circles of Nazi-sympathizing far right out of mere circumstance: there were long-standing factors that played into the decisions he made and the outcomes that resulted. The present work seeks to illustrate those factors, providing an account of how and why a wealthy aristocrat with scientific pretensions and respectable academic qualifications ended up detained for the first half of the Second World War and his career in tatters thereafter.

Indeed, it is this latter point about the role of science that is critical to understanding these circumstances. Pitt-Rivers almost always presented himself exclusively as a scientific researcher, even when he had objectively digressed into the political sphere. It is clear that he believed 'science' supported his views, and he claimed that his opponents were simply unable or unwilling to accept the veracity of his arguments. At the same time, however, there was no inevitability about how these views and activities might end. Though it could be argued that the very title of this book implies that Pitt-Rivers and the Nazis were always intractably bound together, this was not the case. There were many opportunities for Pitt-Rivers to choose a different course, and, as will be shown, many warnings along the way. He had fallings-out with two of his close friends before 1939, both fundamentally over politics. Following his release from detention, one of his relatives commented simply that, 'you must admit you asked for trouble, and unfortunately the authorities took you seriously and you got it'.[16]

Yet Pitt-Rivers was convinced of the righteousness of his views. By the mid-1930s, he saw himself as one of a noble cohort 'working for peace' with Germany. He later claimed that his actions and travels to the country throughout the period were motivated by only the highest expressions of patriotism, even when these involved corresponding with Hitler and his deputies directly. Pitt-Rivers' case befuddled the British Security Service and the Home Office, with an intelligence officer speculating that he was feigning madness to be released from custody. After spending months searching his home, opening his mail and conducting interviews, MI5 agreed that Pitt-Rivers posed no direct threat to national security but had to remain in custody because he might simply be unpredictable and uncontrollable otherwise.

It is worth noting that despite his pretensions and actual accomplishments, Pitt-Rivers has been largely forgotten in existing historical accounts. Ironically, his own story was largely superseded by those around him. His second wife became a Fellow of the Royal Society; his former mistress became a significant political figure and published an extensive autobiography; one of his sons became the renowned anthropologist he himself had hoped to become. Using the uniquely twenty-first-century measure of cultural memory, Wikipedia, the existing myopia becomes clear. At the time of writing, there is no entry for George Pitt-Rivers, but entries are available for his grandfather, his first father-in-law, his second wife, his mistress and two of his three sons.[17]

This state of affairs was partially Pitt-Rivers' own making. Unlike many in his circle of friends, he published no autobiography or memoirs during his life or posthumously. His closest attempt was a small pamphlet entitled *The Manor House of Hinton St Mary* that purports to be a history of his estate but is in fact dominated by his own story.[18] No publisher could be found for the project, probably in part because it contained attacks on Churchill and the British government that were unpalatable in the aftermath of the Second World War, and the few privately printed copies that exist are nearly impossible to obtain. There is no evidence that Pitt-Rivers ever undertook a more substantial autobiography, beyond the fact that he segregated some sets of his correspondence into files labelled 'memoirs'.

Until 2009, these files and his other papers remained in the attic of his former home, where they were found and generously donated to the Churchill Archives Centre, Cambridge, by the Pitt-Rivers family. They are now available to be consulted by researchers and comprise the main documentary base for this book. In addition to these materials, a number of other sources have also played an important role in this project. One of the most exciting moments of researching Pitt-Rivers' life was the discovery that his mistress and companion on his journeys through Germany in the mid-1930s had later become politically prominent in her own right and penned an autobiography. The discovery of these materials has given her an important voice on the events she witnessed and allow her story to be told alongside that of her employer and companion.

In addition to personal papers and memoirs, this project has also utilized a wide cross-section of government documents related to Defence Regulation 18B, the wartime legislation that allowed for the detention without trial of British subjects suspected of posing a threat to national security. At the time of writing, many additional files on this topic remain closed or are otherwise unavailable to researchers, though the National Archives is opening new items on a regular basis. While I have gained access to several files through the Freedom of Information Act in the course of this work, cross reference numbers within the documents suggest that there is a substantial body of material that is not yet available to the public. As a result, historians still know too little about 18B and those who were affected by it. There is a pressing need for further research in this area as new materials become gradually available. In addition, future scholars may well build upon the narrative presented here through the examination of personal papers not yet available alongside government documents.

All this being said, the life of George Pitt-Rivers is thus the story of both achievement and failure. It is an account of a man born at the end of the nineteenth century who tried to cling to the last vestiges of a lost world for his entire life. He was constantly living in the shadow of his grandfather, trying to compete with the memory of a man whom he had hardly known but whose reputation completely overawed him. This backward-looking perspective would take him into the circles of Europe's most radical regimes in the aftermath of the war that ended the world of his youth. Ultimately, it would be this clinging to a bygone era, coupled with his personal prejudices, that would doom his promising academic career and discredit him. To understand his obsession with the past, it is critical to first examine the world he was trying to recover, and for that reason this account must begin with his distinguished grandfather and the world of the nineteenth century.

CHAPTER ONE

A World Destroyed

The morning of 22 February 1881 was grey and drizzly at the Pyramids. For the two British visitors there that morning, the weather must have been reminiscent of their home country. The pair was not travelling together, though they had known each other in England. The younger of the two men – still only in his late twenties – was engaged in a vast and impressive surveying project of the Great Pyramid and the surrounding area at Giza, on which he would publish a book two years later, and he was living in a nearby tomb during his fieldwork.

The older man was travelling to Thebes to investigate flint flakes embedded in the side of ancient tombs and, reaching the Great Pyramid, asked his guide to stop to allow him to measure an angle on the structure. After observing one another for a time, the older man approached the younger, remarking that 'we seem to be working in the same line' and introducing himself as General Augustus Henry Lane Fox Pitt-Rivers, two-time president of the Royal Anthropological Institute (RAI) and one of the most well-known archaeologists in Britain. The men had met before, when the younger had presented a paper to that very body, but the general's 'difference in dress & whiskers' had obscured the identification. General Pitt-Rivers now queried the younger man: 'Do you know anything of Mr Petrie, he's done a good deal in this line.' The younger man reached into his pocket and produced a card with his name – W.M. Flinders Petrie – and the pair shared a laugh about their mutual lack of recognition. After a friendly chat, they parted ways, with the general heading towards the Sphinx and Thebes and Petrie remaining to continue his work. The pair subsequently corresponded until the general's death two decades later. Petrie himself went on to become one of the most famous Egyptologists of the age, holding the first chair in the subject at University College London, publishing a key work on excavation methodology and uncovering ancient cites and sites around Egypt and Palestine until his death in 1942.[1]

On that morning in 1881, however, it was Pitt-Rivers rather than his young friend who enjoyed fame and academic reputation. Since 1851, the general had been a compulsive collector of artefacts – initially firearms and antique locks, later expanding into more exotic ethnographic items – and in the 1860s had begun dabbling in archaeological digging in Ireland and England. In 1880, an inheritance from his great uncle, the 2nd Baron Rivers, gave him a vast private estate replete with archaeological sites to explore and, in 1882, he was appointed the first Inspector of Ancient Monuments tasked with convincing his fellow landowners that ancient sites on their land should be preserved and excavated rather than face deterioration or destruction. As his biographer has observed, General Pitt-Rivers ran in remarkable intellectual circles that included Charles Darwin and his 'bulldog' Thomas Huxley, and Pitt-Rivers himself was an early advocate of Darwinian evolution and the principle of natural selection.[2] Evolution, he believed, was not only a biological phenomenon but also a social one, and because 'nature makes no jumps' in the natural evolutionary process, human society and politics should seek similar gradualism rather than radical change. His collections and exhibitions were designed to illustrate this evolutionary principle – arranging firearms and lock-key sets chronologically to show how they have developed over time, for instance – with the accompanying conservative, anti-revolutionary political implications remaining mostly unstated and for the viewer to conclude.[3]

More than a few of the general's academic connections had come through his extended family. Born in 1827 with the surname Lane Fox, Augustus Henry was the second son of a military officer who died when he was five. His grandfather was James Lane Fox, who had married Marcia Lucy Pitt, the daughter of the first Baron Rivers, making the 2nd Lord Rivers his great uncle. Under normal circumstances this convoluted connection to Lord Rivers – and the fact that he himself was a second son – would have meant that Augustus Henry would inherit little at best.[4] Sensibly, he followed in his father's footsteps and embarked on a military career. In 1841 he entered Sandhurst, where he underwent six months of training and subsequently joined the Grenadier Guards.[5] It was in the military that he first had the opportunity to show his intellectual prowess and scientific inclinations. With the exception of the Rifle Corps, in the middle years of the nineteenth century, the army's standard-issue infantry firearm was still a variation of the smoothbore 'Brown Bess' musket that had been in service since the 1720s. Rifle technology had developed substantially in the meantime and the British army risked falling behind its rivals if it failed to adapt with the times. A committee was established to test various options and make recommendations for the new infantry weapon that should be adopted, and, in 1851, Lane Fox was invited to join the deliberations.[6] The committee eventually recommended the Minié rifle for adoption, and the following year he travelled to Europe to study methods of mass instruction for the new weapon. In 1853, he became the lead instructor at the newly established School of Musketry and implemented an instructional system mostly of his own devising.[7]

By early 1854, trouble was on the horizon. With tensions rising with Russia in the Crimea, the Grenadier Guards were transferred to Malta and Lane Fox soon joined them to supervise training there. In September he sailed with General Sir George de Lacy Evans and landed in the Crimea, taking part in the Battle of the Alma. He was subsequently promoted to major and mentioned in despatches by General Evans. In October, he was invalided home as the result of a long-standing lung ailment but returned to Malta in 1855 to lead rifle instruction training.[8] In 1857, he purchased the rank of Lieutenant Colonel; in 1861, he was transferred to Canada as tensions rose with the United States; and in 1862, he served as Assistant Quartermaster General in Cork, Ireland. After a six-year absence from active military service, he returned to command the West Surrey Brigade Depot in 1873, was promoted to major-general in 1877 and was subsequently given the honorary rank of lieutenant-general. He formally retired from active service in 1882.[9]

In 1853, Lane Fox married Alice Stanley, a daughter of Lord Stanley of Alderley. The Stanleys had long-standing ties to the Grenadier Guards and it is likely that the couple met through relatives moving in the same social circles.[10] The Stanley family was large and distinguished: Alice's sister Katharine became Viscountess Amberley and the mother of Cambridge philosopher Bertrand Russell, while her sister Blanche married the 10th Earl of Airlie and was an ancestor of the Mitford sisters and Clementine Churchill through her daughters. The Stanleys had a reputation for being 'rude, quarrelsome and lively' – Alice was described by her mother as being 'too short', and Lane Fox's first offer of marriage was rejected by her father because his social prospects were seen as insufficient – but after being given permission for the marriage, he appears to have fit in with the family well.[11] Throughout his life, Lane Fox was known for his sometimes-violent temper, domineering attitude towards his family and employees and his devotion to eccentric interests. He was regarded in the family as an atheist and demanded cremation rather than burial, which was unusual for the time and evidently derived from the cremations he had uncovered as an archaeologist. On one occasion, Alice was heard expressing her distaste for the practice to which the general blusteringly replied, 'Damn it woman, you shall burn!' The general was eventually cremated while his wife, who survived him by years, was buried in the village churchyard.[12]

Alice had her own eccentricities. Her nephew, Bertrand Russell, recalled that every time he visited there was never enough to eat and if any food remained on breakfast plates, it was returned to the dish and served later to another diner. The reason was not poverty – the general's finances were in fine shape – but his wife's obsessive frugality.[13] Russell regarded the family as 'quite eccentric' and recounted that on another occasion, Alice had planned a social function at the house, only to find that none of the guests arrived. Investigation subsequently revealed that the gates of the estate had been locked because the general considered the event 'frivolous' and refused to allow the visitors on the property.[14]

In November 1855, Alice gave birth to a son, Alexander Lane Fox. A second son was born the following year, and several girls followed. By 1870, the relationship between Alice and her husband had deteriorated significantly, in part, according to the general's biographer, because she had taken a strong interest in the military career in which he himself had decreasing interest.[15] Since the early 1850s, the general had been creating collections of firearms, locks and other antiquities he had acquired on military deployments. By 1874, the collection was so vast that it filled his London house 'from basement to attic' and as a result, he loaned it to a branch of the South Kensington Museum in Bethnal Green when he departed to resume military service. He made continuous additions to it for years afterwards.[16]

In addition to collecting, the general had another interest as well: archaeological digging. During his time in Ireland in the mid-1860s, he undertook digs on ancient walled settlements known as raths, sending artefacts to the British Museum. After returning to England, he undertook a large-scale study of ancient forts in Sussex, publishing his findings in 1869, and then proceeded to undertake archaeological explorations in Oxfordshire, Holyhead, North Wales and London, where he examined Thames gravel deposits exposed by recent construction.[17] In 1878, he visited Brittany and Denmark, digging in both, and in 1881 visited Egypt and encountered Petrie in the shadow of the Great Pyramid. He subsequently proceeded to Thebes to examine tombs in the Nile Valley and examine the flint flakes embedded in their walls in the hope of showing that they were the result of Palaeolithic human tool use rather than natural deposit. His subsequent publication on the topic was controversial.[18] By the mid-1870s, the general was known as an excellent excavator and an outstanding documentarian of archaeological sites. As a result of his achievements, he was elected to the Royal Society and twice as president of the RAI.[19]

In 1880, the general's prospects and fortunes changed dramatically when Horace Pitt, the 6th Baron Rivers, died without heirs. The 6th Baron Rivers was a cousin of the general and normally this would have meant nothing, except for an unusual clause that had been inserted in his great uncle's will. This specified that the 2nd Baron Rivers' estates and those of another relative, Lord Bingley, should never be inherited by the same person as long as there were two living descendants of Rivers' sister, Marcia Lucy Pitt, between whom the lands could be divided. Marcia Pitt had been the general's grandmother, and, fortuitously for him, his elder cousin had already inherited the Bingley estates on his own. This meant that the general would be given the Rivers estate on the conditions that he prove willing to adopt the surname Pitt-Rivers and the Pitt family coat of arms within one year of the inheritance.[20] This was unexpected good fortune. Having taken the Pitt-Rivers name and the appropriate arms, the general abruptly became the owner of 27,000 acres, making him one of the largest landowners in the country. In an episode of typical petulance, the fact that he lacked a title to accompany these extensive holdings irked him. He subsequently wrote to Prime Ministers William Gladstone and Lord Salisbury asking for the title of Lord Rivers to be granted on the basis of his landholdings rather than merit or achievement. Perhaps predictably, these requests were refused.[21]

With his new financial security, the general could focus his energies on his burgeoning interest in archaeology. He moved from London to the Rushmore manor house in Wiltshire, which had been Lord Rivers' family seat, and also inherited a plush townhouse in Grosvenor Gardens.[22] The Pitt family had moved to Rushmore after an earlier Lord Rivers sold Stratfield Saye, a large stately home in Hampshire, to Parliament, which subsequently presented it to the Duke of Wellington in recognition of his Napoleonic War victories. The current Duke of Wellington retains the house to the present day.[23] Though Rushmore was a significantly less grand home than Stratfield Saye, it had particular appeal for the general. Notably, the estate sits in the Cranborne Chase, a plateau of land that for centuries had the status of a royal hunting preserve. It was said that in the thirteenth century King John had been a frequent visitor to the Chase, and because of the royal control of the land, residents had long been forbidden from interfering with either the wildlife or the forests. The result was that Cranborne Chase had a vast array of almost completely untouched archaeological sites when the general inherited it.[24] For his remaining years, the general was able to dig on his own land – without outside interference – and retain any artefacts he uncovered. Hiring a series of assistants to help with the physical digging and the documentation of the finds, the general opened dozens of burial mounds (barrows) that had never before been explored and dug up a number of Roman fortifications and other sites.[25] In 1889, he led the investigation and restoration of a house in the village of Tollard Royal that was rumoured to have been King John's hunting lodge. Stripping away centuries of modifications and making methodical records of his discoveries, the general uncovered and restored medieval features consistent with a thirteenth-century structure.[26]

The general was now in the last decade of his life but showed little sign of slowing his activities despite declining health. Relations with his family remained strained: his eldest son, Alexander, was completely dominated by his father and was probably a disappointment to him. Ironically, he would also be seen as a disappointment by his son.[27] The second son, St George, had technical inclinations and invented a prototype incandescent light bulb before Thomas Edison. Several of the general's other sons showed some interest in his archaeological work but had few of his talents, with the exception of their artistic abilities. His daughters were hardly happier. The oldest married a Stanley and the second married Baron Avebury, a Liberal politician who was a friend of Charles Darwin and served as one of his pallbearers. Lord Avebury was himself a keen archaeologist and preservationist who had worked with the general in these pursuits.[28] The general's youngest daughter, Agnes, entered literary circles and become an author and supporter of women's suffrage. She was posthumously immortalized by the author Thomas Hardy, a frequent visitor to Rushmore, in a commemorative poem entitled 'Concerning Agnes' that recalled them dancing together at the estate on an August evening.[29]

Despite, or perhaps because of, his familial strife, the general pushed on with his work until nearly his last days. His obsession with putting his findings into the public sphere as an educative measure led to his most lasting achievement. In 1884, he donated his collection from the South Kensington Museum, which by now contained thousands of items from all over the world, to Oxford University. The legal stipulations of the gift were twofold: firstly, a building had to be constructed to house the collection and it could be used for no other purpose and, secondly, a lecturer had to be appointed to teach the subjects covered by the collections. These conditions were met with the construction of the Pitt-Rivers Museum, Oxford, which remains to the present day, and the appointment of the first anthropology lecturer in the country. When Oxford agreed to these terms, the collection was transferred and the Pitt-Rivers Museum established to house it.[30]

This was a remarkable achievement on its own, but the general regretted one aspect of the arrangement: he would no longer have control over the collection and its display. This fact irritated him, particularly when he inevitably argued with the anthropology lecturer, E.B. Tylor, and the museum curator, Henry Balfour.[31] As a result, the general founded a second museum in Farnham, Dorset, where he deposited additional objects, including a number of valuable artistic works from Benin. Unlike its Oxford counterpart, the Pitt-Rivers Museum in Farnham and the collections it housed would remain the family's private possession.[32]

The general's final years were occupied by several major tasks. In 1882, he had been appointed Inspector of Ancient Monuments, a new position created in large part through the efforts of Lord Avebury. The Ancient Monuments Act of 1882 provided for the hiring of an inspector to travel the country examining ancient monuments and making a list of those recommended for preservation by various means, including the erection of fences around them to prevent damage. The act was toothless, however, and if a landowner refused to have a site listed, there was no legal mechanism to challenge their decision. The general was an obvious choice for the position: he was a major landowner, giving him credibility with the wealthy squires he would encounter, and was also one of the leading experts on ancient sites in the country. He diligently travelled the country with his assistants, documenting nearly 100 sites by 1890, but few actually received government protection.

By the time of the general's death, there were only forty-three officially protected monuments in the country. The owners of the remaining sites had refused to allow the government to intervene.[33]

In addition to his work on ancient monuments, the general committed himself further to the idea of public education. Both the Pitt-Rivers Museum in Oxford and the Farnham museum were impressive achievements, and he saw the latter as particularly important to his mission. Many of the exhibits in Farnham were models or reproductions of objects held elsewhere, as the general believed that an 'educational' museum should not compete with 'research' museums holding the originals for expert study.[34] In addition to the museum, the general constructed a nearby pleasure garden to attract visitors who might be tempted to step into the building and learn during their visit. Called the Larmer Tree Grounds, the general designed and envisioned a space in which families from far-flung villages in the countryside could meet and avoid loneliness, which he believed was a cause of madness.[35] The grounds were more than just gardens and included bandstands, a theatre, a temple and various buildings in a variety of styles from around the world. The general even hired his own band to perform at the grounds on Sunday afternoons – scandalizing some in the religious community – and the musicians wore the traditional blue and gold livery colours of the Pitt-Rivers family.[36] The band did not last long, evidently in part because the players were paid little and there were episodes of drunkenness, but the grounds were remarkably popular and attracted nearly 45,000 visitors in 1899.[37] At one stage, it even included a small zoo of exotic animals. This was not only for the enjoyment of visitors but also the general himself, who took an interest in animal husbandry and at one point attempted to crossbreed his yaks with various varieties of cows and oxen.[38]

The general died in 1900 and was remembered as one of the leading figures of nineteenth-century archaeology. His meticulous record-keeping during field excavations has led him to be praised as the 'father of scientific archaeology'.[39] There is no doubt that he was one of the first excavators to adopt and consider the principles of stratigraphy – the idea that the section of the earth in which an artefact is found is significant – and while his excavation techniques were not in line with modern standards, he was an innovator in his time.[40] His belief in arranging artefacts chronologically to demonstrate their evolution – and in his view, the gradual evolution of human society, without revolutionary leaps – demonstrated the importance of archaeological objects and sequences for the study of anthropology. The two museums that were founded by the general and arranged on these principles stood as a memorial to his accomplishments and ideas.[41] As Debbie Challis has recently observed, the general's influence extended deep into the key intellectual and scientific circles of the period. In the 1880s, he sat on committees headed by Flinders Petrie and Francis Galton, the founder of the eugenics movement, that were tasked with racially classifying ancient civilizations based on archaeological evidence.[42] The general's direct connections to Galton – and his interest in racial classification – would later be sources of pride for his grandson.

Within the general's family, however, all was not well. His children quarrelled over the terms of his will and Alexander's management of the estate, forcing the Rushmore house to temporarily be let to a tenant. There was litigation over the Larmer Tree Grounds and the Farnham museum, with some claiming that the latter belonged to a charitable trust rather than the family. This argument was eventually rejected, and the family's ownership of the museum was confirmed.[43] By the time of the general's death, Alexander also had a burgeoning family of his own. In 1889, he had married Alice Ruth Hermione Thynne, the daughter of Conservative politician Lord Henry Thynne. Henry Thynne's father had been the 3rd Marquess of Bath, his elder brother would become the 4th Marquess, and his nephew the 5th Marquess. The Thynnes were primarily politicians and soldiers, and Alice Ruth's brother Ulric joined the King's Royal Rifle Corps and would be mentioned in despatches in both the Boer War and the First World War. While his father was alive and living at Rushmore, Alexander and his family resided at a manor house in the Dorset village of Hinton St Mary. The house and surrounding lands had been part of the general's inheritance from Lord Rivers but had been let for decades to various tenants. In 1888, the general decided that he desired the house for his own purposes and evicted the renters, who were by then in their late 80s, over their piteous protestations that they had resided there for years and wished to stay until the end of their lives. The general subsequently installed his eldest son and his family in the house.[44]

On 22 May 1890, Alice Ruth gave birth to the couple's first and only son, George Henry Lane Fox Pitt-Rivers, in London. A girl, Marcia Ruth Georgiana, followed in late 1891. In contrast with the general's large family, they had no further children.[45] By this time, Alexander had embarked on a brief career, serving in the Royal Wiltshire Yeomanry and the Dorset Yeomanry. As a young man, he had trained as an architect and worked as a clerk for the architect designing the Royal Courts of Justice in London. He was evidently a skilled watercolour painter.[46] Artistic ability aside, however, he was nowhere near as talented or distinguished as his father. While he showed 'an interest in his father's antiquarian pursuits', he was primarily occupied with estate management matters and county responsibilities as Justice of the Peace and High Sheriff of the County. George Pitt-Rivers would later mourn his father's life as 'solitary, isolated, uneventful, monotonous and without achievement'.[47]

Relatively little is known about George Pitt-Rivers' early life. He met his distinguished grandfather on numerous occasions as a young child, but the general's death when he was just 10 years old robbed him of the opportunity to know him as an adult. Pitt-Rivers' obsession throughout his life with his grandfather assuredly resulted from the fact that his knowledge of him was almost exclusively secondary and the realization that he would have had far more in common with the general than his own father. The memory and legacy of his famous forebear hung over nearly everything he wrote, and his frequent references to the general's archaeological and anthropological work in both personal correspondence and published writings far outnumber references to his own father and mother. George Pitt-Rivers' own son subsequently noted that his father viewed the general as 'the idealised father-figure throughout his life' because he was 'totally antipathetic to his parents'.[48]

Following the general's death and the resolution of the conflicts related to his estate, Alexander moved his family from Hinton St Mary to Rushmore. Life there was reflective of the upper-class gentility of the Edwardian Age, with frequent balls, dances and dinner parties that included the leading political and social figures of the period. The house was full of antiquities collected by the general and other works of art, making a striking impression on the guests who came to stay and visit. In 1908 Sir Almeric FitzRoy, long-time clerk of the Privy Council and a social acquaintance of the Pitt-Rivers and Thynne families, visited the house and was charmed by both the estate and his hosts:

> At Rushmore, where the Pitt-Rivers's [sic] have recently installed themselves, we enjoyed a hospitable shelter.... The beauty of the situation, on a spur of the downs at the edge of Cranborne Chase, and the antiquarian interests with which the neighbourhood was endowed by the researches of the late eccentric owner, provide eye and mind with plentiful objects of pursuit. Mrs. Pitt-Rivers is a delightful hostess of the type that relies largely on Providence, with a very bright and intelligent little daughter. The view from the top of the down covers the whole coast from the Needles to the Isle of Purbeck, and to the north along the great folds of Salisbury Plain with its western termination in the wooded slopes that are spread beneath Stourton tower. The glades of the [Cranborne] Chase are cut so as to show its manifold beauties, and the abrupt undulations of the ground give them the mystery of some fantastic maze.[49]

This charmed high-society life continued at Rushmore until the outbreak of war. For the young George Pitt-Rivers, however, it would be interrupted sooner. He was sent to Eton and left the school in 1909. He appears to have been a relatively undistinguished student, winning no major recognitions or scholarships in his time there.[50] One of the few surviving pieces from his early life is a short essay written at Eton in 1908, at the age of 18, on the topic 'How far socialistic ideals are practicable?' As the formulation of the title itself suggests, his essay argued that the answer was not positive.[51] Like his grandfather, his health was questionable at times and he suffered from asthma throughout his life.

After leaving Eton, Pitt-Rivers followed in family tradition and embarked on a military career by initially joining the Royal Wiltshire Yeomanry. Rather than eventually joining the Grenadier Guards in which his grandfather had served, he later joined the 1st Royal Dragoons, a renowned and famous cavalry regiment that had distinguished itself at the Battle of Waterloo by capturing a French infantry division's colours. His first choice unit had been the 9th Lancers, of which several relatives were members, but he was denied entry because he had not attended Sandhurst and had no real military experience. He joined the Royals as a 2nd Lieutenant and would be promoted to Lieutenant in 1913. His first deployment was to the city of Muttra in northern India, where he arrived to join the regiment in 1910.[52] The area was relatively peaceful during his time there, and most of his time was taken up with training and drills.

In July 1911, Pitt-Rivers obtained leave to return to Rushmore for a celebration of his twenty-first birthday. The public festivities were a reflection of the charmed life at the estate in the pre-war period, extending nearly a week and including thousands of local residents and children.[53] The first portion of the merriments took place at Hinton St Mary, where local residents were invited to visit the manor's extensive gardens while their children were entertained with games and 'a first-class conjurer'.[54] The following afternoon, 1,500 guests descended on Rushmore for a luxurious celebration that included tea in the gardens, music from a military band and games for the children in attendance. The accompanying dinner lasted for several hours and included the presentation of an antique silver salver to the young Pitt-Rivers on behalf of his father's tenants and various speeches honouring the heir to the estate. Pitt-Rivers was articulate but unpolished in his response to these encomiums, already demonstrating the beginnings of the oratorical skills that he would use to great effect later in life. As a local newspaper reported:

> Mr. George Pitt-Rivers was heartily cheered in rising to reply. He said he was sure that whatever words he might say would be absolutely inadequate to express his real feelings. He had no words at his command to tell them how he would value the beautiful present they had given him. He might say he valued it exceedingly, not only for itself but even more for the sentiments which were attached to it. (Applause.) He only hoped that the feelings of goodwill, which he believed had long existed on that estate between landlord and tenants – and he knew he was expressing his father's sentiments – might long continue – (applause) – and that whatever contingencies might arise, whatever the actual legal relationship between landlord and tenant might be, nothing would abate the feeling of goodwill which existed now (Applause.) He would be returning to India in a few days, and he could only say that the most pleasing memory he would take back with him would be that proudest hour of his life. (Applause.) Whatever he could say was inadequate to the occasion,

so they must forgive him saying those few words in such a lame style ('No, no.') He hoped they all realized that he would say much more if he could, but they would understand all he wanted to say. (Loud applause, and musical honours.)[55]

Rising to reply on behalf of the tenants, a representative proposed a toast to 'the squire's health' and expressed a hope that 'Mr. Pitt-Rivers and his family would never take to themselves the idea of putting their estate into the market like so many of the landlords round them were doing, because he was convinced they were better off as tenants under Mr. Pitt-Rivers than as buyers of farms (Hear, hear.).'[56]

Following the speeches and toasts, a large fireworks display took place in the manor's gardens that included 'a clever production of a portrait of the heir to the estate, with the words "Long life and happiness"'. This impressive feat was 'received with the greatest enthusiasm' by the crowd. As the guests began to disperse, some were returned to their home villages by a special train that had been procured for the purpose.[57] These glamorous celebrations and reassuring declarations of social harmony would soon belong to the world of the past. This was still fundamentally the late Victorian world of his grandfather, with its clearly defined social distinctions and roles. Just a few years later, the gentility would begin to crumble.

In the meantime, however, Pitt-Rivers himself had little time to enjoy the remaining vestiges of the pre-war world. In November 1911, the Royal Dragoons were sent to South Africa and stationed near Johannesburg.[58] The presence of the Royals was not coincidental: since the end of the Boer Wars, South Africa had been gripped by a series of gradually escalating labour disputes that threatened the stability of the state and the profits of its leading industrialists. One of the hotspots was the mining industry, which had long been a centre of disputes over working hours, safety measures and compensation.[59] In 1907, Jan Smuts, then serving as colonial secretary and in charge of the Transvaal region, deployed troops to suppress a mining strike and protect the replacement 'scabs' brought in by the management to replace picketing workers.[60] Smuts' action helped set the precedent for state intervention to protect privately owned mines and other interests from labour actions.

After the formation of the Union of South Africa in 1910, a national military was created in the form of the Union Defence Force, but it was slow to be established and the country's security still relied on the British troops stationed there. In the event of major unrest, the South African government had little choice but to call upon the troops available to them. Trouble came in mid-1913 when a serious labour dispute broke out at a mine in the Transvaal. Following a change in working hours intended to increase profits, five mechanics quit their posts and their union instructed members not to fill their positions. The management subsequently appointed non-union workers to replace them, leading to a strike among the remaining employees. Refusing to negotiate, the management then brought in 'scabs' to replace the striking workers, but they were assaulted by the picketers, and workers at other mines soon joined the industrial action in solidarity. Smuts, by now serving in multiple positions that included the powerful combination of both minister of defence and minister of finance, was unable to negotiate an end to the strike and instead prepared to crush it with force. Calling up nearly 3,000 imperial troops to assist the 2,800 police and 1,600 special constables already deployed, the government prepared for a major confrontation.[61]

The Royal Dragoons were soon sent to the centre of the emerging conflict in Johannesburg. As a mounted unit, the Royals could move quickly through the streets and use their horses for crowd control purposes in the event of a riot. On 4 July, a rally of 5,000 strikers and sympathizers took place at the central Market Square and became unruly after a number of inflammatory harangues by speakers. The Royals were ordered to disperse the crowd, which responded by throwing stones and bottles. At least one soldier was knocked unconscious in the melee. The rioters then set fire to a nearby train station after overwhelming the police garrison protecting it and opened fire with revolvers on the Royals and the police when they attempted to intervene. The police subsequently returned fire with rifles, killing several rioters, while the Royals charged the crowd and struck them with the flats of their swords. Demonstrators subsequently attacked the headquarters of several mining companies and burned down the offices of a newspaper that was considered a mouthpiece for the owners. The mob then drove away the fire brigades that attempted to put out the flames and armed itself by looting a gunsmith's shop.[62]

Fighting continued into the night, as 'the troops repeatedly charged, and the crowd consistently scattered and re-formed … for an hour after midnight, there was incessant firing in the riot area, and repeated cavalry charges and volleys'.[63] Dynamite was thrown by the rioters, which the Royals believed now consisted mostly of 'hooligans' and 'brutes' rather than striking miners, and shots were fired by both sides.[64] The following morning, Prime Minister Louis Botha and Smuts arrived in the city, further inflaming the situation and leading to the police firing on a crowd threatening the posh Rand Club. By the afternoon of 5 July, more than two dozen civilians lay dead, with hundreds more rioters, bystanders, police and soldiers wounded. Botha and Smuts hastily signed a deal with the union leaders reinstating all striking workers to their previous positions and promising a full investigation of their grievances. Amidst the violence, the strikers had effectively won.[65]

The Royals, and Pitt-Rivers, had been at the centre of this vicious confrontation. 'No more distasteful experience than this month's strike-duty at Johannesburg can have ever come the way of a British cavalry regiment', the official history of the unit remarked.[66] The civilian casualties that resulted were horrific but might have been much worse if the Royals had been less disciplined in their use of force. Many of the dead had been shot by the police rather than the cavalry and some had been trampled by the mob in the stampede to set buildings alight.[67] A subsequent judicial enquiry found that the Royals had used only necessary force under difficult circumstances, and their actions had probably saved lives on both sides. Pitt-Rivers' exact role and experience in the confrontation is not documented, but as a junior officer he would have been expected to take a leading role. There is no doubt that he was involved in the melee in some capacity: in the immediate aftermath, he took photographs of the damaged buildings, the burned out newspaper office and dead Royal Dragoons horses that had resulted from the violence.[68] It is likely that the experience further hardened his views towards socialism and communism that he had already expressed at Eton. If rioting and unrest on such a scale as this could result from a mining strike, what horrors might come with a larger-scale uprising?

Smuts himself learned a similar lesson from the 1913 riots. When the white railwaymen's union threatened to trigger a potentially catastrophic general strike in January 1914, the government immediately placed protection around key buildings and railway depots. Key union leaders were pre-emptively arrested and weapons sales were temporarily suspended in potentially volatile areas. More than 10,000 South African soldiers and police were deployed before the strike was even officially called, and when it was formally declared, the government immediately proclaimed martial law. Elite units surrounded the hall that served as the strike headquarters, and when they pointed a piece of field artillery at the building, the strikers inside surrendered. Nine of the leaders were subsequently taken in the middle of the night and forcibly deported from the country. Though they returned just months later, the South African Parliament retroactively approved Smuts' order for the deportations, effectively legitimizing an expansive use of martial law to combat industrial unrest.[69] The traumatic experience of 1913 had convinced Smuts that strikes and disorder had to be met with overwhelming force. Pitt-Rivers himself would largely agree in his later writings.

The Royals spent the last months before the Great War attending to more tranquil duties. They were moved from Johannesburg to the smaller city of Potchefstroom, where they had been stationed before the 1913 unrest. Pitt-Rivers was promoted to Lieutenant and much of the winter and spring was consumed with drills and musketry training. By summer, the fragile peace that had held Europe together for nearly a century was crumbling. The unification of Germany in 1871 had upset the traditional balance of power on the continent, and the Balkans had already been in a state of increasing unrest for years. On 28 June 1914, Archduke Franz Ferdinand and his wife were fatally shot as they rode in an open-top car in Sarajevo. The Serbian gunman, Gavrilo Princip, was not acting alone and was part of a larger conspiracy to assassinate the heir to the Austro-Hungarian throne. Ethnic violence against Serbs resulted immediately across Bosnia, and Austria-Hungary issued a punitive ultimatum to Serbia that it expected to be refused. Behind the scenes, Austria-Hungary asked the German Kaiser, Wilhelm II, if the country would honour its treaty obligations and enter a war against Serbia. The Kaiser agreed and when the Serbian government rejected the ultimatum, Austro-Hungarian troops bombarded Belgrade on the night of 28 July.[70]

Serbia had allies of her own. On 24 July, the country had appealed to the Russian Tsar, Nicholas II, for assistance in the event of war. The Tsar agreed, and on 31 July mobilized troops to the country's entire western border, including the frontier with Germany. Faced with the potential for a devastating invasion from the east, Germany demanded that Russia stand down immediately. When the Tsar refused, Germany declared war on 1 August. Russia was also not without allies, having signed a pact with France in 1892. Western Europe would now be dragged into the war, and when attempts at mediation were rejected, Germany declared war against France on 3 August and marched into neutral Belgium the following day. Britain had pledged to defend Belgium's neutrality in the 1839 Treaty of London and, while there was debate about the actual treaty obligations that were still relevant, London issued an ultimatum demanding complete German withdrawal from the country. When Germany did not comply, Britain declared war on 4 August.[71] Watching night fall over London in the hours before the ultimatum's expiration and the coming of war, Foreign Secretary Sir Edward Grey later recalled remarking to a friend that 'The lamps are going out all over Europe, we shall not see them lit again in our lifetime.'[72]

The lamps had indeed gone out remarkably quickly, and Britain was now involved in its largest-scale conflict since the Napoleonic Wars of the early nineteenth century. Nation-wide mobilization took place immediately and reservists were called up. Thousands of young men stood in queues to join the military and by the end of August, more than 100,000 had enlisted.[73] Regardless of their feelings about the war itself, crowds turned up at railway stations around the country to support the soldiers heading towards battle.[74] At Eton, students nearing the end of their studies lined up to join, as did many Old Etonians. As horrendous casualty figures and death counts made their way back to the school and its alumni over the coming months, one student recalled his older brother, who was shipping out imminently, fatalistically remarking, 'Don't worry about this. We are *all* going to get killed. You and I and everybody else.' Four months later, the older brother lay dead in France.[75]

South Africa was far from the action, which disappointed some members of the Royals who were eager to experience the adventure and excitement of a large-scale war. 'It is horrible being tucked away here at a time like this', one member of the unit wrote after the outbreak of war.[76] They did not have long to wait for their share of combat, however. When war was declared, the South African government announced that it no longer required the assistance of British troops, releasing those units for service elsewhere. The Royals immediately prepared to depart for home and left the country on 23 August. In late September, the regiment arrived in Southampton and prepared to mobilize for European service.[77]

The Royals had already missed out on the first dramatic weeks of the war while they were in transit from South Africa. Plunging into Belgium in the opening days of the war, the German army encountered heavy resistance from the Belgian army at Liege before smashing through in the middle of August and crossing the French border. The British Expeditionary Force (BEF) encountered the German army near Mons in Belgium and was soon forced to retreat across the border to the south. The defeat demoralized the British, who were more used to fighting under-gunned natives around the Empire than well-armed and trained European armies. By early September, the German army had suffered heavy casualties but crossed the Marne River. Paris itself was now potentially in jeopardy, and a desperate and bloody struggle developed. At one stage of the battle, legend holds, French troops were rushed to the front in Paris taxicabs. The Germans were eventually forced to withdraw and the capital was temporarily saved.[78] The only flanking opportunity for either side now lay to the north, and the Germans focused on trying to outmanoeuvre the allies there. The British and the French moved divisions to counter the German advance in what became known as the 'race to the sea', and in mid-October the armies clashed near the strategically important city of Ypres in Belgium.[79]

The First Battle of Ypres was another desperate and bloody struggle that would leave more than 250,000 men dead or wounded in a matter of weeks.[80] The Royals had now fully mobilized and were ready to join the conflict. On 7 October, the regiment was transported across the English Channel to the port of Zeebrugge. From there, they marched and rode towards Ypres to link up with the main line of the BEF. Arriving on 14 October, the Royals encountered detachments of the German army, but these were only small-scale skirmishes. One frustrated officer wrote that the war thus far had consisted of 'marching and counter marching' with no discernible objective.[81] The boredom ended when the Royals were ordered to advance towards German-held villages to the south and encountered heavy resistance. Forced to retreat, a Royals squadron was pursued down a road with deep canals on both sides and nearly pinned down by German rifle fire. It only escaped annihilation thanks to the appearance of a British armoured car with a mounted machine gun that could cover the withdrawal, along with the purportedly poor marksmanship of the German soldiers.[82] It was a narrow and lucky escape for the men involved, but the overall situation would soon get worse. The following morning, the Royals were sent to reinforce an entrenched position in the French line. Dismounted from their horses, which were rapidly being killed, they dug trenches under a heavy artillery barrage before being pushed back. This heralded a new tactic for the Royals: they would now be used as a mobile reserve, riding on horseback to reinforce weakened portions of the line and then dismounting to fight in the trenches. It was an effective strategy but would increase their casualties dramatically.[83]

The Royals were now playing a key role in their sector. With German attacks intensifying, their new role as mobile reserve provided important reinforcements to units that were pinned down and suffering heavy casualties. For days, the Royals rushed to various problem spots, dismounting from their horses to fight. 'The only thing that now distinguished the cavalry from the infantry was that they had no entrenching tools or bayonets and, being Corps reserve, might be rushed anywhere at any moment', the official regimental history noted.[84] In early November, they were once again ordered into the trenches under shelling so heavy it gradually buried the men alive under dirt and debris if they could not dig themselves out. Heavy rain flooded the trenches and German snipers picked off those who dared to go aboveground.[85] On 17 November, the Germans unleashed a massive artillery bombardment against the Royals' position and followed it up with a large-scale infantry attack. More than 400 attackers were mown down by British machine gun and rifle fire until 'the ground in front [of the trench] was thickly covered with German dead'.[86] This major attack thus repelled, the British divisions were withdrawn from the line and replaced with French troops. Despite being used as an infantry reserve rather than as a cavalry unit, the Royals had distinguished themselves.[87]

There had been a cost to this heroism. Since arriving in Belgium six weeks before the Royals had taken nearly 100 casualties, and the dead included a major, two captains, a second lieutenant and twenty-three men.[88] Another nine officers had been wounded. This latter category included Pitt-Rivers, who had been wounded by German machine gun fire in the onslaught of the seventeenth. The bullet struck his left leg just below the knee, shattering the bone and ripping apart his muscles. It would take years for the wound to close, and his left leg and foot would never properly function again. Years later, he would have to lean on the arm of a friend to navigate the busy streets of Paris and by the end of his life, he would carry a heavy walking stick.[89]

Returned to England for surgery and recuperation, Pitt-Rivers' war was effectively over. He would return to the trenches at the end of the conflict on light duties but his injuries were too extensive for a return to front-line duty. Convalescing from his injury, he turned his attention to reading and writing. He would soon publish his first short book after studying the works of psychoanalysts Sigmund Freud and Carl Jung. There was more to do at home than just work, however. In December 1915 he married Rachel Forster, the daughter of politician Henry William Forster, Conservative member of Parliament for Sevenoaks and financial secretary to the War Office. It was a wise choice for an ambitious young man with extensive political connections of his own and a family background in both the

Liberal and Conservative parties. Rachel was attractive and charming, and Almeric FitzRoy described her as having 'rare gifts of thought and language, which she handles with a graceful diffidence beyond all praise'.[90] She had a passion for acting and the theatre, which she would soon adopt as her primary interest. In 1917 she gave birth to the couple's first son, Michael, and in 1919 a second son, Julian, followed.

Pitt-Rivers' connection to the Forster family would soon prove advantageous to his career, though his marriage with Rachel would rapidly turn sour. He was already aware that the pre-war world of Rushmore gentility, the Larmer Tree Grounds, and his grandfather had been lost for good in the bloodshed and unrest of the war. As the conflict ground on year after year, his cynicism grew. By the end of the war, Pitt-Rivers had convinced himself that Europe had placed itself on an even more dangerous precipice than it had sat on in the summer of 1914. With ample time to read, write and reflect, he soon claimed to have found the source of the world's ills and committed himself to its neutralization. The fact that he would purportedly make this discovery through his study of German philosopher Friedrich Nietzsche would have far-reaching consequences.

CHAPTER TWO

From Nietzsche to the Antipodes

The eventual announcement of the November 1918 Armistice and the subsequent Treaty of Versailles officially ending the First World War sparked huge celebrations across Britain. David Lloyd George's wartime Liberal-Conservative coalition was subsequently given a fresh mandate at the polls, and with the 'War to End All Wars' now over, there was hope for a lasting peace among many. For many ex-servicemen, including George Pitt-Rivers, however, the experience of conflict had convinced them that war itself could never again be allowed to take place. This urgent drive to avoid a second world conflict would soon lead some in dangerous directions.[1]

For Pitt-Rivers, the end of the war meant the end of the most formative experience of his life. Initially holding the romantic notions of war that were common to many young men of his generation, by the end of the conflict, he had become sceptical of its aims and consequences. He returned to the Western Front in the final months of the war, taking with him a draft of new recruits and serving on only light duties due to his wound. Writing home from the trenches in 1918, he reported that few soldiers in the field believed the 'colossal humbug' he had heard back in Britain and that the 'undying hatred and the shrieks of vengeance increase the further you go behind the line, until they reach the climax of hysterical rage in London, Paris and Berlin ... the heroics of those heroes at home who are all for fighting to the last drop of somebody else's blood, are not universally appreciated out here'.[2] He would later denounce the 'political war-mongers in England' who he believed were responsible for both the outbreak and the bloody continuance of the war.[3]

In many senses, Pitt-Rivers had had a 'good war'. The wound he suffered in November 1914 was physically devastating and would become nearly debilitating in later years, but it may well have saved his life. As the official history of the Royal Dragoons noted, Pitt-Rivers had been among the first British soldiers to encounter the Germans in the area east of Ypres in 1914 and there were few other regiments that had taken as comparatively few casualties under the circumstances. At the same time, the battle had been a 'grievous trial' for the men, who had faced shelling from three sides in November 1914 and fought for several straight days with the realization that there were few infantry reserves behind the front line they were defending.[4]

This would prove to be only a prelude of what was to come for the Royals, however. Casualties mounted as they were repeatedly ordered to dismount from their horses and enter front-line trenches. As long as the Royals remained on horseback, the unit had a degree of mobility and could manoeuvre quickly as a mobile reserve; dismounted and sent into the trenches, they were merely highly trained infantry. Following a reshuffle of officers in mid-1915 due to the resultant death toll, Pitt-Rivers was promoted to adjutant of another unit, the Lothian Horse and Yeomanry, though he was still 'too lame from his wound for active service but fit for home duty'.[5] He was eventually ordered to rejoin his reserve unit despite the severity of his injuries, at which point, he later claimed, he 'ignored orders' and entered a nursing home to undergo a second operation on his leg.[6] In June 1917 he was promoted to Captain, giving him the title that he would hold for the rest of his life. He was subsequently given a payment for 'the value of a partial permanent disability of one leg' of about £356 (the equivalent of approximately £15,000 in 2015), which he later spent building a wall between the Hinton St Mary manor house and the churchyard next door.[7]

Many members of the Royals were far less fortunate. By the Armistice, the unit had lost twenty-four officers and 148 men of other ranks. Another twenty-five were wounded. As the regimental history noted, while these numbers were small in absolute terms in comparison to some infantry units that lost thousands of men, the Royals had not expanded greatly in size during the war, making the unit's proportional casualties high. In more direct terms, a total of 108 officers had served with the regiment from 1914 to 1918: of them, a full 45 per cent had been killed or wounded by the Armistice. Of the original thirty-four officers listed on regimental rolls in August 1914, among whom Pitt-Rivers was included, 35 per cent were dead by the end of the war and most of the others had been wounded at least once and 'several of them more than once'.[8] Few Royals emerged from the war unscarred, if they lived to see its end at all.

Despite Pitt-Rivers' honourable service, however, the post-war world initially held little promise for his military career. In August 1919, the Royals returned to England after briefly serving as an occupying force on the Rhine. With the war concluded, the unit concentrated on rebuilding from its losses with officers who had been holding staff appointments elsewhere re-joining the unit. This influx of returning officers meant that

Pitt-Rivers' chances of further promotion were low, particularly as his still-healing leg wound might imperil his ability to be deployed to the field. In addition, his early wounding meant that he had little actual combat experience and there were officers with far more time in combat and heroism under fire who would be promoted before him. Facing few prospects for advancement under these circumstances, Pitt-Rivers retired from the active list of officers in 1919, citing his father's declining health and the need for his 'assistance in the management of a large agricultural property'.[9] It was, in some senses, good timing: in April 1920, the Royals were deployed to Ireland to fight the Irish Republican Army in the midst of 'very unpleasant conditions' that claimed additional lives.[10] Despite no longer being in active service, Pitt-Rivers would remain on the list of reserve officers for the unit until his final resignation in late 1936.

When peace came in 1918, Pitt-Rivers believed Europe had been destroyed, both physically and spiritually, and now sat on the brink of an abyss. The late-Victorian world of his father and grandfather, with virtually unchallenged British mastery of the seas, the power and resources of the empire, and the political security of the Concert of Europe, now lay in ruins. After the war, he feared, there would have to be a new path for Western civilization to take if it were to avoid collapse in the face of the most pernicious enemy it had ever faced: Bolshevik Russia and international communism. Just as revolution had swept Russia in 1917 and led to the establishment of the world's first lasting socialist state, the fall of Britain's enemy, the German Kaiser, at the end of the war was not to be celebrated but feared, he wrote, because it 'must inevitably culminate in a world outbreak which no civilized country will be able to withstand...[because] the change from Democracy to Bolshevism (contrary to the opinion of those who will not see it) is swift and easy'.[11] This quest for a new philosophical path to lead the world from the grip of communism would soon take Pitt-Rivers into circles discussing a radical new philosophy in the Anglophone world: the ideas of German philosopher Friedrich Nietzsche.

Nietzsche himself had died in Saxony in 1900 after a lengthy illness that had precluded extensive work for nearly a decade. At the time of his death, there was still comparatively little understanding of his ideas and, indeed, a number of his significant works had not yet even been published. In the English-speaking world he was virtually unknown except among a few German-speaking experts and enthusiasts.[12] In essence, Nietzsche's writings argued for the radical overcoming of traditional ethical constraints and the teachings of Christianity, both of which he referred to as 'slave morality'. Some individuals would be able to overcome these traditional views and achieve true greatness by creating their own laws and values, he argued, becoming an *Übermensch* (a super human). Following the 'death of God' in the modern world, he argued, individuals and societies now had to create their own foundations to build upon, and they would need an aristocracy of the great and the good – individuals who had overcome the slave morality – to lead them. But how could these ideas be practically implemented in the political sphere? Nietzsche was a philosopher, not a politician, and those convinced by his ideas were left to their own devices when it came to practical implementation. The debate over Nietzsche's ideas and intentions would occupy philosophers throughout the twentieth century and beyond, but this was still far in the future and at the turn of the century, his work was known by only a few experts.

By 1907, however, this had begun to change with the American publication of H.L. Mencken's *The Philosophy of Friedrich Nietzsche*, a book that presented a cross-section of his philosophical ideas to English-speaking audiences along with a translation of one of his shorter works.[13] Among the British converts Nietzsche attracted before the First World War was A.R. Orage, a socialist who had been introduced to Nietzsche's book *Thus Spake Zarathustra* in 1900 and was immediately 'reborn and rejoicing in the Nietzschean gospel'.[14] By 1907, Orage was publishing short tracts introducing Nietzsche's ideas to the public and, according to a biographer, his biggest consideration was now 'whether Nietzsche was the whole truth – or the truth for him'.[15]

For Orage, Nietzsche represented a completely new way of viewing society and social progress. In contrast to the widespread Victorian optimism that had marked the nineteenth century, Nietzsche 'was a force threatening to destroy all faith in progress and liberal idealism. He was therefore a portent, prophetic of the end of an epoch', Orage claimed.[16] The new era that would emerge, Orage believed, would be a combination of socialism with Nietzsche's idea of an *Übermensch* able to radically overcome existing social mores and conventions to create a new morality that could guide the world forward. In 1907, Orage took these ideas to the public by purchasing *The New Age*, a radical London-based periodical that he quickly transformed into a leading venue for *avant garde* ideas and discussions. Though he himself maintained strongly socialist views, as editor he believed he should 'write as little as possible, but concentrate his energies upon the inspiration of his team and the organization of their labours'.[17]

This did not mean, however, that Orage's views would go unrepresented, and he often made space available in the publication for his fellow Nietzscheans. Among his regular contributors in the pre-war years was Oscar Levy, a complicated figure whose role in early twentieth-century Anglophone intellectual circles was extensive and is still largely under-examined by historians.[18] Born in 1867 to a Jewish family in Pomerania, Levy decided to leave his family's banking business to become a medical practitioner. In 1894 he settled in London, later claiming that his move was at least in part motivated by the increasing imperialism of Germany.[19] He was soon introduced to Nietzsche's works by a female patient and claimed to have experienced an epiphany during a visit to the British Museum when he realized that 'Monotheism etc. may not be "Progress" after all'.[20] Over the coming decades, Levy devoted his life to the furtherance of Nietschezean thought in the Anglophone world. In 1906 he published a short book entitled *The Revival of Aristocracy* that posited the future emergence of a new Napoleon that would be 'destined to lead mankind out of democracy and Christianity to realms higher and brighter'.[21] In 1909, he published the first four volumes of what would become an eighteen-volume translation of Nietzsche's complete set of published writings, along with a comprehensive index. While Levy did not translate every volume personally, he oversaw every aspect of the project with an almost missionary zeal, as he reflected to readers of *The New Age* on the project's completion:

> While we may well be modest about what we have done, it would be absurd to play the humble hypocrite about the fact that we have done it, that we have been able to secure a public for Nietzsche in England at all. For England was no doubt the most important country of all to conquer for Nietzschean thought. I do not mean on account of her ubiquitous language … I am thinking of another and more important reason, which became a conviction to me during the progress of this publication: the firm conviction that if we could not obtain a hearing for Nietzsche in England, his wonderful and at the same time very practical thought might be lost for ever to the world – a world that would then quickly be darkened over again by the ever-threatening clouds of obscurantism and barbarism.[22]

In addition to his profound belief in the power and importance of Nietzsche's ideas, he held another view that would endear him to Pitt-Rivers: anti-Semitism. In Levy's reading of Nietzsche, Jews were responsible for the creation of the decadent and decaying values system that had led to the development of Christianity – a form of 'Super-Semitism' and a cult of weakness – and, ultimately, the First World War and the cataclysms that had followed.[23] These 'Jewish' values stood in opposition to the new philosophy that Levy believed had been revealed by Nietzsche, and therefore Jews represented the status quo that would eventually be overcome:

> The world still needs Israel, for the world is in a period of quick change and requires a centre round which the best of all nations may rally and recover. The world still needs Israel, for the world has fallen a prey to democracy and needs the example of a people which has always acted contrary to democracy, which has always upheld the principle of race. The world still needs Israel, for terrible wars, of which the present one [the First World War] is only the beginning, are in store for it …. The world still needs Israel, for the old values of Israel are being severely

questioned, and no one knows any longer what is good and what is evil…. I therefore sincerely, and contrary to Nietzsche, hope that the Jews will refuse to be insorbed [sic] and absorbed, and that, if they should waver, the anti-Semitic 'bawler' should come on and drive them back to their duty.[24]

Levy openly accepted the label of anti-Semite but saw his own views as vastly different from 'vulgar' anti-Semites who preached simple anti-Jewish bigotry. He believed that his own anti-Semitism was influenced by the 'enlightened' ideas of Nietzsche and, as a result, offered a revolution in thought rather than simply prejudice.[25] By the end of the First World War, Levy had become a prominent figure in the modernist circles surrounding *The New Age* and between 1907 and 1920 he published more than sixty essays in the publication, oversaw the Nietzsche translation project and published longer works of his own, making him one of the leading lights of the early English language Nietzschean movement.[26]

Embarking on his own quest for post-war understanding, Pitt-Rivers soon entered these circles and his connection with Orage and Levy would change the course of his career. Following his resignation from military service, Pitt-Rivers began to pursue academic interests and in January 1920 arrived in Worcester College, Oxford, as a Fellow Commoner studying psychology and social anthropology. He remained there for two terms and his military service was granted as substitute for three term's study, giving him nearly enough terms of study to quality for a degree.[27] He briefly studied under psychology professor William McDougall – an expert on evolutionary psychology with an interest in eugenics – before McDougall's departure for Harvard in 1920. Pitt-Rivers' main academic interests were the ideas and writings of Sigmund Freud, the philosophical writings of Nietzsche and his anthropological studies. These combined in 1919 when he published his first short book, a short volume entitled *Conscience and Fanaticism: An Essay on Moral Values*, before even beginning his formal studies.[28]

Purporting to be a philosophical examination of human ethical judgements, the thin book reads more like an extended undergraduate essay than a work of significant scholarship. Indeed, the publisher requested that Pitt-Rivers pay part of the production costs, as 'I do not think the very interesting MS. [manuscript]…is likely to meet with sufficient success at the present moment to encourage me to take it up in the ordinary way of business.'[29] 'Interesting' is perhaps the most generous conclusion one could draw about the text. The work began with a discussion of 'moral intuitions' and denunciations of organized religion ('priestcraft') and conscientious objectors in times of war but quickly became a confused jumble of ideas and arguments culminating in, as one reviewer put it, 'a rather confusing result' through which 'incidentally we learn that he condemns conscientious objectors, Russian Bolshevists, loud-voiced newspapers that boast of controlling public opinion, and [Catholic writer] Mr. Hilaire Belloc'.[30]

The overall conclusion of the work was that human 'intuitions' about right and wrong based in religion are fallacious and lead individuals only to 'fanaticism'. Individuals should instead make moral judgements derived from John Stuart Mill's ideas of utilitarian value (the most good for the most number of people), coupled with the idea that the nearest utility ('proximity') has the most value (meaning that a decision benefitting individuals nearby takes precedence over a decision that might create more good a further distance away). Even if individuals themselves do not see the benefit, or utility, of a decision, he wrote, they have a psychological understanding of it through their heredity and the subconscious.

As a result, Pitt-Rivers argued, moral judgements are derived from what individuals think will create the most immediate good for those closest to them, however they might define that group. The fact that societies and nations generally perceive these interests similarly can lead to 'mass suggestion' or the convincing of an entire group that a given action should be undertaken even if it does not appear rational.[31] The actions that a group will consider undertaking is again conditioned by heredity and the unconscious, so different social groups will view the same ideas in divergent ways.[32] A truly effective demagogue, he argued, understands what it takes to convince 'the masses' that a course of action is desirable even when it is not.

It would be an understatement to say that *Conscience and Fanaticism* was a bad book: it was a pastiche of ideas that had hurriedly been patched together during his convalescence from war wounds. It is best read as both an immature work and also the writing of a man recovering from very real mental and physical wounds. Its negative references to conscientious objectors and religious notions of morality were reflections of his own disillusionment with British society and the war. However, these themes were lost in the hodgepodge. The few readers and reviewers who managed to make it to the end of the book reported having little idea what point the author was trying to convey beyond a general dislike for religion, pacifists and the press ('organized concerns for the propagation of lies and counter-lies').[33] Even the kindest reviews were scathing. A writer in *The British Medical Journal* acidly concluded his review by refusing to pass a final comment on the book due to Pitt-Rivers' stated distrust of the press.[34] Both commercially and in terms of reception, the book was a failure. Pitt-Rivers himself rarely mentioned it in his later career, listing it with his other publications but never appearing to place much value on it. Virtually everything else he would publish would gain greater success and recognition.

However, *Conscience and Fanaticism* bears mention for several important reasons. Firstly, Pitt-Rivers' emphasis on inherited group judgements has significance for his later work, particularly regarding race. If ethical understandings of right and wrong were passed through the generations and affected by past and present social conditions, groups would develop homogeneous views of right and wrong that affected their judgements to the present. Going a step further, as Pitt-Rivers would, and assuming that these social groups were equivalent to different races, one could argue that each race possessed innate behavioural traits that could never be conditioned away (a communal conscience, or what he called 'Cosmic Suggestion'). The notion that these views affected individuals' judgements at the subconscious level would make the situation even more pernicious and ensure that people might not even be able to know why they make certain decisions. In essence, according to Pitt-Rivers, each group (race) had a perspective on the world that was inherited at the subconscious level and could never be overcome. This notion of rigid racial inheritance of not only traits but also mental states would become a trademark of his later work.

In addition, Pitt-Rivers also used the book as a platform to put forth the ideas that would later take him into the eugenics movement. Echoing Freud's views about human violence being rooted in sexual repression and pressure to conform natural instincts to social norms – themes explored in Freud's 1913 book *Totem and Taboo* and later in *Civilization and Its Discontents* – Pitt-Rivers argued that the outbreak of war in 1914 had been caused by a combination of psychological factors and the general increase in the numbers of 'unfit' individuals in the population.[35] These groups were being allowed to breed and thrive in large numbers, he claimed, due to the negative influences of religion and democracy:

> The great world war has indeed emphasized the immense power of ideas. We hear much of propaganda and ideals. In medicine we hear more of 'psychotherapy', or the treatment of disease by persuasive and hypnotic methods…I do not suggest that the causal origin of the European War is purely psychic in character, it may with greater certainty be found years before its disastrous developments, in the steadily increasing pressure of population, assisted by the gradual elimination of the natural checks among the indigent and unfit and the proportionate increase in the burdens of the fit, due chiefly to the growth of democratic ideas and trend of religious influences; this pressure found expression in policies of expansion among the more prolific nations, and in the case of Germany, where relief could not adequately be found in colonization, as a natural consequence engendered assiduous military and bellicose propaganda, which was bound eventually to culminate in a world war.[36]

This was a crude view of human society that attributed the outbreak of the war to the rise of the 'unfit' and the supposedly increasing influence of 'religious ideas'. While borrowing heavily from the trendy psychoanalytic ideas of the day, this took the idea of psychological complexes precipitating violence to an extreme. Had Germany been given access to colonial areas where the country's 'unfit' could be sent, he was suggesting, the war might have been avoidable.

Despite its clear shortcomings and controversial conclusions, *Conscience and Fanaticism* marked Pitt-Rivers' emergence on the intellectual stage and, more significantly, attracted the attention of *The New Age*, which published a lukewarm review by Nietzsche enthusiast and translator Anthony Ludovici.[37] Pitt-Rivers responded to Ludovici's critiques in print and in subsequent letters to the editor, marking the first time he had been invited to contribute to the publication.[38] With this new collaboration established, in mid-1920 Pitt-Rivers made his intentions to publish a new, more ambitious work known to Orage, Ludovici and their colleagues at *The New Age*. This new book would directly integrate Nietzsche's ideas about the dangers of Christian morality, coupled with Levy's ideas about the need for a robust aristocracy, to analyse recent world events, specifically the 1917 Russian Revolution. Orage immediately suggested that Pitt-Rivers talk to Levy about his project, and in May the two men met for the first time at Pitt-Rivers' club in Central London. It was the beginning of a friendship that would last until the end of Levy's life and one that would radically affect both men's careers.

The theme of Levy and Pitt-Rivers' first meeting was summed up in one phrase: 'we must act – *now*'.[39] Both men shared the view that 'we must have a new religion', based in Nietzsche, to combat the rise of communism and that there was not a moment to spare before their attack should be launched. Pitt-Rivers' new literary project was intended to put forth precisely these views, with a secondary agenda as well: to expose what he claimed were the 'Jewish' origins of the Russian Revolution. By enlisting Levy's help, both men hoped the book would be spared the criticism of being anti-Semitic and be given a higher level of credibility. As Levy himself told Pitt-Rivers, 'you *may* be called an Antisemite [*sic*], but if so, I, the Semite, may be branded with the same name'.[40] Both men would indeed soon face this label.

Levy and Pitt-Rivers soon reached an agreement about how their collaboration would proceed: Pitt-Rivers would secure a publisher for the manuscript he had been working on already and Levy would write an extended 'prefatory letter' to accompany it. The book would be provocatively entitled *The World Significance of the Russian Revolution* to attract a wide readership and emphasize its far-reaching topicality.[41] Levy believed that his letter would not only help Pitt-Rivers avoid criticism of his anti-Semitic views but also, bizarrely, be beneficial to the Jewish community itself. The writing of such provocative sentiments 'had to be done', he told Pitt-Rivers, and 'of course my brothers in race will hate me for it, because I have attacked their three favourite schemes of to-day [*sic*]: yet much is the ransom of straightforward literature. But I am an old hand at the game and don't mind being unpopular'. By doing so, he claimed, the issues raised would be elevated from 'the Anti- and Philosemitic' to a 'higher and healthier plane'.[42]

If it was truly Levy's hope to use his contribution to elevate the tone and reception of the book, his plan backfired magnificently. As was typical within his Nietschze-influenced outlook, Levy's preface argued that Jews had been responsible for virtually every historical development in Western history, specifically the introduction of Christian morality and culture that had culminated in the catastrophe of the First World War and the rise of communism.[43] 'You have noticed with alarm that the Jewish elements provide the driving forces for both communism and capitalism, for the material as well as the spiritual ruin of this world', he told readers.[44] By promising a 'new Heaven' through the subsequent development of Judaism, Christianity, capitalism, communism, and all other political and spiritual creeds, Jews had led the world into a 'new Hell' devoid of moral progress, he claimed. Even worse, the architects of this destruction, under the influence of Zionists, now sought to flee into a future Jewish state in the Middle East to escape the results of their misconduct.[45] According to Levy, however, Jews had a moral obligation to remain in Europe and right their previous wrongs with the help of Pitt-Rivers and his fellow anti-Semites, who were in fact the best friends of the Jewish people:

> Yes, there is hope, my friend, for we are still here, our last word is not yet spoken, our last deed is not yet done, our last revolution is not yet made. This last Revolution…will be the revolution against the revolutionaries….The great day of reckoning is near. It will pass a judgment upon our ancient faith, and it will lay the foundation to a new religion…then you, my dear Pitt-Rivers, the descendant of an old and distinguished Gentile family, may be assured to find by your side, and as your faithful ally, at least one member of that Jewish Race, which has fought with such fatal success upon all the spiritual battlefields of Europe.[46]

In many ways Levy's prefatory letter, which stretched more than a dozen pages, was a summary of the entire short book and, set in the same type size as the rest of the book, might well have matched Pitt-Rivers' forty-four-page contribution in length. Denouncing Vladimir Lenin as a 'neurotic' and the Bolsheviks generally as 'crazy paranoiacs and parasites', Pitt-Rivers argued that the Communists had managed to defeat the Tsar's forces, 'because no movement representing a heterogeneous jumble of contradictory and incompatible elements can ever defeat another movement which at any rate knows its own mind and allows of no compromise'.[47] In other words, the Bolsheviks had been able to seize power because they had the uncompromising will to do so. Their opponents 'were far too "democratic" to be either disciplined or efficient'.[48]

Thus far, this was a fairly standard argument that bears the clear hallmarks of Pitt-Rivers' interest in Nietzsche and Levy-esque belief in the need for strong aristocratic leadership to lead the world. However, halfway through the text the argument deviated in a strongly anti-Semitic direction. The Russian Revolution, Pitt-Rivers claimed, had been precipitated by an international Jewish conspiracy. There were very few actual Russians among the Bolshevik leaders, he claimed, and 'It would be futile to suggest that all Bolsheviks are Jews, or that it is *chiefly* a Jew movement. I am only trying to show why, as a matter of *fact*, Revolutionaries are so largely Jews in Russia.'[49] This emphasis on presenting what were in reality unsupportable suppositions as 'fact' would become one of Pitt-Rivers' hallmarks in his later years, and it is instructive that even at this early stage his technique was well developed.

The bulk of his argument for Jewish responsibility for the Revolution was presented as follows:

> It is not unnaturally claimed by Western Jews that Russian Jewry, as a whole, is most bitterly opposed to Bolshevism. Now although there is a great measure of truth in this claim, since the prominent Bolsheviks, who are preponderantly Jewish, do not belong to the orthodox Jewish Church, it is yet possible, without laying oneself open to the charge of anti-semitism [*sic*], to point to the obvious fact that Jewry, *as a whole*, has, consciously *or unconsciously*, worked and promoted an international, economic, material despotism, which, with Puritanism as an ally, has tended in an ever-increasing degree to crush national and spiritual values out of existence and substitute the ugly and deadening machinery of finance and factory. It is also a fact that Jewry, as a whole, strove every nerve to secure and heartily approved of the overthrow of the Russian monarchy, which they regarded as their most formidable obstacle in the path of their ambitions and business pursuits. All this may be admitted, as well as the plea that, individually or collectively, most Jews may heartily detest the Bolshevik regime. Yet it is still true that the whole weight of Jewry was in the revolutionary scales against the Tzar's [*sic*] government. It is true their apostate brethren, who are now riding in the seat of power, may have exceeded their orders; that is disconcerting, but it does not alter the fact. It may be that the Jews, often the victims of their own idealism, have always been instrumental in bringing about the events they most heartily disapprove of; that maybe is the curse of the Wandering Jew.[50]

The argument was, in essence, that Jews were opposed to the Tsar's government, either consciously or unconsciously, and were therefore responsible for the victory of the Bolsheviks, even if they themselves, individually or as a group, opposed communism. Any action taken by the Bolsheviks was therefore the responsibility of all Jews, regardless of their personal views and actions.

What evidence was there for these startling claims? Pitt-Rivers claimed that they were simply objective facts, available to anyone willing to look:

> Certainly it is from the Jews themselves that we learn most about the Jews. It is possible that only a Jew can understand a Jew. Nay, more, it maybe that only a Jew, can save us from the Jews. A Jew who is great

enough, strong enough – for greater racial purity is a source of strength in the rare and the great – and inspired enough to overcome in himself the life-destructive vices of his own race. It was a Jew who said, 'Wars are the Jews' harvest.' But no harvests so rich as civil wars.[51]

In light of this argument, Levy's preface takes a different context. As a Jew himself, it is clear that Pitt-Rivers believed him to be uniquely positioned to reveal the 'truth' of his wider claims. The psychological claim that all Jews 'consciously or unconsciously' sought these outcomes is significant as well: even if an individual Jewish person was not directly involved in the revolution, they still shared responsibility for the outcome due to their unconscious mental states. In addition, communism itself was 'psychologically, economically and morally' a return to primitive society – a 'lapsing back' – that could only be brought about by the destruction of traditional family structures.[52] In this effort, he claimed, the Bolsheviks were trying to 'communize women' by making them 'the property of the whole nation' rather than husbands. All Russian women were therefore being subjected to mass rape under the Bolshevik system, he claimed.[53]

With these bewildering arguments, the book's concluding paragraph was offered a prediction of the bloodshed that would follow two decades later:

> The pages of history are blackened with the records of the misery and suffering men have created for themselves, of countless human holocausts as horrible and senseless as those of Dahomey [an African kingdom renowned for human sacrifice], whose tortured victims have been destroyed – self-immolated, for the most part – on the altars raised to vain words and meaningless symbols. And still their crazy priests and fanatical votaries, mad with frenzy and drunk with blood, shriek for ever more victims, never content until the whole world is infected with their madness and rocks helpless in an orgy of self-destruction.[54]

While this chilling description may now call to mind the terrors of Nazism and the Holocaust, in Pitt-Rivers' mind this bloodshed would follow in Russia and Europe-wide unless a mass anti-Semitic and anti-Bolshevik consciousness was awakened to fight the threat of Jewish-led communism.

Pitt-Rivers and Levy were fully aware of the furore their collaboration was likely to produce but they did not fully anticipate the consequences. The backlash was slow to begin: Levy complained in September 1920 that the book had 'fallen flat' due in part to a lack of coverage in the major British papers and a 'conspiracy of silence' in the Jewish press.[55] 'They [the Jews] are, I think, afraid of drawing the attention to my preface, which preaches antisemitism [*sic*] from above, and that is a kind of antisemitism they are entirely unfamiliar with', Levy told his friend. 'So there we are … up against the wall of fear, prejudice, cowardice, misunderstanding, as usual. But, pray, do not be disheartened. We have done good work. Time will redress matters.

This very night I dreamt that I had a real fight with Zionists in the streets of London and I remember that I hugely enjoyed it: so I cannot have come out "second best".[56]

Just months later, responses to the book had become nearly overwhelming. The first critical review appeared in *The Morning Post* in early September, and Levy was surprised that a number of early reviewers focused their ire on his preface as much as on Pitt-Rivers' portion of the text.[57] Most writers were critical: *The Times Literary Supplement*, for instance, concluded that Pitt-Rivers possessed 'a rich store of the language of invective' but that the book had few facts and little interesting to say. 'His animosity against Bolshevism seems only a little stronger than his animosity against his own country', the review concluded.[58] A letter writer in the right-leaning magazine *The Spectator* denounced Levy's contribution as 'rodomontade', while another attacked Pitt-Rivers and Levy for engaging in 'the ancient and, one had hoped, discredited pastime of *Jew baiting*. It certainly comes as a shock to the student of men and manners to notice that, in this year of grace, this medieval sport should still be in vogue'.[59] Pitt-Rivers' claim about women being 'nationalized' under the Bolsheviks was roundly mocked in the *Birmingham Gazette*, which found Levy's introduction to be 'brilliant, but highly-debatable'.[60] *The Spectator*'s own reviewer mused that Pitt-Rivers' claims about Russian women was 'a curious point, and one which we must return to on some future occasion' and called the book 'thoughtful and suggestive' overall.[61]

Anti-Semites were more impressed. Among others, the book's ideas appealed to Nietzsche's rabidly racist sister Elisabeth Förster-Nietzsche, who was apparently moved to tears by Levy's preface, praised Pitt-Rivers' argumentation and asked for several copies of the book.[62] At least one Conservative Party candidate for Parliament claimed to carry a copy of the book in his pocket at all times, and American writer and Nietzsche commentator William M. Salter fawned over the book in a letter to Levy.[63] Industrialist Henry Ford's anti-Semitic newspaper *The Dearborn Independent* published an adulatory review praising both Levy and Pitt-Rivers under the headline 'Dr. Levy, a Jew, Admits His People's Error' and describing the book as the result of 'unprejudiced observation'. It also reprinted much of Levy's preface verbatim.[64] Anti-Semite John H. Clarke similarly wrote that Levy's introduction had confirmed the veracity of the notorious 'Protocols of the Learned Elders of Zion', a purported secret plan for Jewish world domination that had been published in English in the early twentieth century and was widely (and rightly) known to be a plagiarized forgery.[65] As Dan Stone has noted, Levy was unhappy with many of these reviews, which he believed had misunderstood his argument and associated him with the 'vulgar' anti-Semitism he saw himself opposing.[66]

In the midst of this unfolding controversy, a major problem presented itself when Pitt-Rivers abruptly announced that he was leaving the country.

His father-in-law, Henry Forster, had been appointed to the Privy Council in 1917 and was created Baron Forster of Lepe in 1919. In June 1920, he was named Governor General of Australia and quickly offered Pitt-Rivers the position of Private Secretary and Aide-de-Camp in Melbourne. With Pitt-Rivers' military career over, this was potentially the opportunity of a lifetime for him. With evidently little hesitation he accepted the offer and began preparing to leave England. Almeric FitzRoy, the old social acquaintance of the Pitt-Rivers and Forster families, was deeply saddened when he heard the news of the Forster family's planned departure. 'Farewells, particularly at my time of life, are touched with melancholy, and I cannot see without emotion such delightful personalities pass out of my view', he wrote. 'Never will Australia have received a richer gift from the Mother-country than in these three examples [the Forster family] of the fine flower of English life in its most impressive significance.'[67]

For Levy, this was more than simply a sad moment and bordered on disaster. He was stricken when he heard the news, remarking to Ludovici that 'it is a serious blow for us'.[68] It was perhaps more of a personal blow than a professional one, because at the same time Levy was involved in a battle with the Home Office over the status of his permission to remain in the country. As a former German subject, Levy was denied long-term residence in the United Kingdom by the 1919 Aliens Act that prevented the settlement of former 'enemies' in the country. The fact that Levy had already lived in the country for years made no difference to his case nor did the fact that the German government no longer considered him a citizen though he evidently still travelled on a German passport.[69] In April 1920, Levy had been granted a visa by the Home Office on the grounds that he needed to conduct urgent business with his publisher but he was given only temporary leave to remain in the country. By September, when Pitt-Rivers had already left the country, the Home Office refused to issue Levy another visa, citing the Aliens Act as the reason. Levy's subsequent appeals on the grounds that he was suffering from a serious heart condition were rejected.

Levy's plight gained widespread publicity in both the press and British literary circles. The question presented by many commentators centred on whether he had been singled out for deportation because of his controversial work on Nietzsche or his recent collaboration with Pitt-Rivers. The rabidly anti-Semitic periodical *The Hidden Hand* blamed 'the learned elders of Zion' for arranging the expulsion of 'the most courageous and honest Jew living', citing the prefatory letter as evidence of Levy's importance.[70] Pitt-Rivers attempted to intervene on his friend's behalf, writing from Australia to inquire about the Home Office's position and inquiring about whether 'there is some other reason for the attitude of your Dept. not hitherto disclosed' for not allowing Levy to remain in England as 'a distinguished and loyal author, with a record of fifteen years residence'.[71] The clear implication of this question was whether Levy was facing expulsion because of his involvement with Pitt-Rivers and the recent controversy surrounding their joint publication. 'Dr. Levy's case is unique, and cannot justly be appreciated if categorized with other unnaturalized Germans... I believe his case may have been prejudices by the ignorance or misunderstanding of the unsophisticated who have communicated to Scotland Yard, denouncing him as a man whose writings were dangerous?' Pitt-Rivers argued.[72]

Despite sympathetic press reports, a small-scale letter writing campaign to the Home Office, the arrival of a petition signed by Levy's Bloomsbury neighbours, and a brief inquiry by the King about the case, Levy was forced to leave England in October 1921.[73] He appears to have been uncertain about whether he was being targeted by the authorities or not: in March he reported to Pitt-Rivers that 'all' his friends believed he was being forced from the country due to his political views and writings on Nietzsche.[74] Surviving Home Office files suggest that this supposition was incorrect and that the government's interest stemmed almost exclusively from his status under the Aliens Act. As a Home Office official summarized it for the King, 'the previous sections of the Aliens Act makes it impossible to allow Dr. Levy to remain longer in this country. He came for a short visit: that visit has been inordinately extended, and now an effort is being made to turn the visit into permanent residence, which would be directly contradictory to the provisions of the Act'.[75] At no stage does the Home Office appear to have considered Levy's work to be a particular threat to national security, and the only intervention from the covert realm came with a brief note from the Director of Intelligence noting that an 'English journalist' had reported Levy as 'a suspicious person'.[76] The Home Office itself questioned whether this intelligence was 'well founded', particularly as 'he is now collaborating with Captain Pitt Rivers upon a critical examination of Bolshevism'.[77] Special Branch of the Metropolitan Police reported that one of Levy's former housemates (possibly the same journalist) had denounced him as 'a German spy' and that the Nietzsche works he had translated 'were, in fact, written by the Kaiser himself for propaganda purposes'.[78] This outlandish tip may well have been the source for the Director of Intelligence's information but it appears to have not been taken seriously in the debate over whether Levy should be removed from the country.

In contrast to the view that Levy's writings had led to his trouble with the government, it appears more likely that his work with Pitt-Rivers had actually increased his standing in the English literary world. Certainly his contribution to *The World Significance of the Russian Revolution* was widely discussed: in protesting his removal from the country, the periodical *Jewish World* remarked that its objections were based in the fact that Levy had not been granted a legal hearing on the matter, 'not because he is of the Jewish race.... Because as a matter of fact his interest in matters Jewish has never been robust, while only a little while ago, in the preface to a book ("The World Significance of the Russian Revolution", by Captain Pitt-Rivers), Dr. Levy expressed an opinion concerning Jews which, to say the least, was as antipathetic as it was untrue'.[79]

Under these circumstances it is unlikely that Pitt-Rivers would have been able to do more for Levy even if he had still been in England. However, the fact that both men had left the country by the end of 1921 removed much of the wind from the sails of *The World Significance of the Russian Revolution*. Had he remained in the country, Pitt-Rivers almost certainly would have promoted the book more widely through his personal networks and continued literary output. For his part, Levy settled in the Rhineland of western Germany, which remained under French occupation, and continued to work, awaiting the next stage in the world-changing Nietzschean transformation that he expected at any time. 'I can only say that nothing has changed, except for the worse', he told Pitt-Rivers. 'But that is the same all over the world. I am consequently anxiously awaiting the second "push" of the world revolution which will make an end to parliamentarians, democrats, lawyers, journalists, jury-women and other "progressive" animals'.[80] He would soon discover a world leader – Benito Mussolini – that seemed to deliver this Nietschzean dream, at least initially.

For Pitt-Rivers, the transition from author to government official was more difficult than he had anticipated. His new position with the governor general placed a new and uncomfortable demand for political silence on him, and in May 1922 Lord Forster asked Pitt-Rivers to change the title of a talk he was scheduled to deliver on communism, as 'the title I suggested [was] too provocative, and too apt to raise political controversy. I must, of course, avoid topics that bear directly on the politics of the country'. The talk went forward under a title suggesting that it dealt primarily with 'primitive Communism and its relation to modern developments in Russia and elsewhere' and carrying a footnote stating that 'the discussion will bear upon the historical and psychological aspects of the subject, and will be non-political'.[81]

Pitt-Rivers would soon find his new political straightjacket intolerable and seek a new venue to put forth his ideas. The vehicle he would choose, anthropology, would allow him to pass many of the same commentaries he had already set forth in a way that attracted less attention from the uninitiated while giving him the academic credibility he craved. Accordingly, by mid-1921 Pitt-Rivers had all but abandoned his actual duties in Australia to pursue the research that would bring him his highest level of recognition and praise.

CHAPTER THREE

Australia and the Science of Man

The year 1920 had been a good one for George Pitt-Rivers. At just 30 years old, the publication of *The World Significance of the Russian Revolution* had given him the publicity, if not the academic acceptance, that he coveted. Yet, his recent move to Australia in the role of Aide-de-Camp to his father-in-law, the governor general, put him in an awkward position. How could he continue his avowed anti-communist and anti-Semitic advocacy when he could not directly discuss politics? This was the dilemma he now faced, and he would soon revolt against its strictures. Ironically, the solution that he found to his predicament would eventually gain him the academic status he desired.

In the meantime, Pitt-Rivers had already left a reasonably significant legacy behind in England when he departed in mid-1920. His 1919 publication of *Conscience and Fanaticism* had been panned in the press but gave him attention from the critics in the literary community surrounding *The New Age*. This in itself was a significant accomplishment, and the fact that he had subsequently attracted a figure as prominent as Oscar Levy to write the preface of *The World Significance of the Russian Revolution* demonstrated a strong potential for future work. On the other hand, Levy had now been forced to leave the country and neither man could reasonably hope to promote the book under their present circumstances. It would have been a good guess that their collaborative effort would have been largely forgotten to all but the most dedicated historians of the era except for later events that cast it into prominence. This, however, was still far in the future in 1920 as Pitt-Rivers sat in the Government House, Melbourne, pondering his future and using his free time to train polo ponies on the ample lawns surrounding the building.

Pitt-Rivers immediately found Australia appalling. His official work kept him 'quite busy' at the expense of the interests 'nearest my heart', leading him to begin plotting travels to 'devote myself to reading, writing, seeing, thinking and, I hope, also understanding'. Even the physical landscape disgusted him: Australian buildings, he told Levy, were poor copies of those in the great cities of Europe 'and invariably a bad copy – more pretentious, more garishly ornate, but constructed of cheap and shoddy material' and Australian houses were 'as a rule the ugliest I have ever seen in any country'. Australians, he went on, insisted on imitating British architecture not out of devotion to the mother country but because they believed 'there is nothing that they can do in England which he can't do "a darned side better"' yet these attempts futilely ended by producing vulgarity. The Australians themselves were 'brave, independent and combative: theirs is the psychology of the hunting pack … but it must be confessed they are as immature as children, wonderfully indifferent to literature or art, an easy prey to demagogues, and very obtuse'. The lack of an elite in Australia was a serious threat, he told Levy, because, 'I feel very strongly that true Aristocracy, which Nietzsche preached, is the only real alternative and the rightful antithesis to Bolshevism and Communism, and that we ought to preach this new hope as an antidote to Lenin's gospel of despair and Sadism.'[1]

Among the few things he seems to have been favourably impressed by was the art of Australian painter George Lambert, with whom he struck up a friendship. Lambert subsequently sketched Pitt-Rivers in pencil and made a bronze sculpture of his favourite polo pony. Momentarily fancying himself an art critic, in 1924, Pitt-Rivers published a book chapter praising his style and artistic achievements.[2] After Lambert's death in 1930, Pitt-Rivers purchased a large oil on canvas painting entitled *The Squatter's Daughter* and had it shipped to England, where it remained for decades. It now hangs in the National Gallery of Australia after being purchased for the nation and returned in the 1990s.

Pitt-Rivers' vocal dislike for Australia and his obvious interest in outside activities suggest that he had little interest in the position he had been given in the country's administration. Lord Forster himself had been chosen as a compromise governor general by Australia's political leaders over two other candidates and, as a result, he was expected to be effectively a non-partisan representative of the Crown, in some ways representing the shifting nature of the governor general's position to its modern role. In practice, this meant that his role was confined to ceremonial and social duties. A consummate sportsman and cricketer, Forster took an interest in yachting and travelled around the country constantly, dedicating a number of memorials to Australia's war dead along the way. His wife committed herself to patronizing music, the arts and the consideration of social questions.[3] Politics was far from their primary interest.

Despite this genteel life within the Government House, Australia itself appeared to sit on the brink of serious unrest in late 1920. In December, race riots broke out in Broome after a Japanese resident was murdered by white residents and both groups formed vigilante groups. The prime minister, Billy Hughes, eventually sent a warship to restore order and apprehend those responsible for the violence. Hughes then concluded that the Australian Communist Party had in part encouraged the riots and in early 1921, the government placed a ban on the import of literature supporting the overthrow of the state. Fear of revolution had fully arrived in Australia, as it had elsewhere.[4] The First Red Scare in the United States had already resulted in the Palmer Raids against leftist groups and the deportation of hundreds of suspected radicals. Communist revolution had swept through Hungary, where a short-lived socialist state was crushed by Romanian troops in 1919. A socialist republic had been declared in Bavaria the same year, resulting in bloody street fighting and its destruction at the hands of the army and the volunteer *Freikorps* (Free Corps). Communism's spread had temporarily been halted at the borders of the Soviet Union, but in this climate there was no telling where revolution might rear its head next.

It was maddening for Pitt-Rivers to have to sit this fight out. In Britain, after all, he was now a recognized anti-communist writer. *The World Significance* had 'done quite well' in sales and the publisher brought out a second edition in Australia in early 1921 in hopes that its success could be expanded and, possibly, foreign-language translations even be secured.[5] Pitt-Rivers himself continued to vigorously defend the claims he had made in *The World Significance* to the greatest degree that he could even after arriving in Australia. In January 1921, he penned a lengthy riposte to the editor of *The Morning Post*, a conservative paper that had panned the book, with its writer privately comparing Levy's introduction to 'a bottle of salad-oil spilt over its pages' and suggesting that Pitt-Rivers might be a Soviet agent seeking to discredit more reputable anti-communists by presenting preposterous arguments.[6] Ironically, *The Morning Post* had only recently published controversial excerpts derived from *The Protocols of the Learned Elders of Zion*, so the anti-Semitic orientation of the paper was hardly in question.

Pitt-Rivers' response letter itself was hardly a demonstration of restraint. Attempting to clarify his and Levy's views, Pitt-Rivers stated that his anti-Semitism 'includes in the semitic [*sic*] movement many Christians and Gentiles, and has on its side many noble-minded Jews', thus directly attacking Christianity along with Judaism. In addition, he stated, the Jews were not 'exclusively' responsible for the Russian Revolution but were 'a factor only, though not the least important one, in the World Revolution'. The real danger posed by the Jews, he concluded, 'lies in their psychology, their distinctiveness, their racial self-consciousness, and separation from the land they live on, rather than the doctrines they assimilate and develop, whether they happen to be financiers or Communists, or neither'.[7]

This was essentially a concise rehashing of Pitt-Rivers' arguments in *The World Significance* but with a slightly new orientation: he was now becoming increasingly interested in the role of psychology and psychoanalysis. Evidence of these could already be found in *Conscience and Fanaticism*, but following his formal studies of the topic at Oxford, he clearly viewed it as a critically important component of his understanding of the world. His studies of Sigmund Freud and Carl Jung's psychoanalytic theories had convinced him that different races possess intrinsically different mental worlds that were effectively unbridgeable and, as a result, races could never truly hope to coexist: hence the importance of his argument that all Jews, 'consciously or unconsciously', bore responsibility for the Russian Revolution.

Pitt-Rivers set out to develop this theory further by contacting Jung himself, sending a copy of *The World Significance* and inquiring whether Jung used different psychoanalytic methods on Jewish and non-Jewish patients. After a lengthy delay, Pitt-Rivers received a reply directly from Jung seemingly endorsing his theory of Jewish psychological difference:

> Your statements concerning the different method of treatment with [Jews] are more or less correct. I say 'more or less', because there is a common humanity underneath the racial differences, which should not be omitted, else you never reach your patient. But it's a long way to that deep layer, where things are equal with Jews. All the upper psychic strata are peculiarly different...they are far more domesticated and far less barbaric than we are, far more subtle and shrewd and in a way much better adapted to various forms of civilization, than we are. They are always aware of the defects of our ideals and of our resources. Thus they particularly grow up on our moral weaknesses and upon our idealistic naivety...
>
> I was really very glad to see how you bring the first of the psychological differences into the discussion. It is an unpardonable unscientific negligence of our time to believe, that we could treat Jews on our own level. I don't say they are worse or better than we are, they are simply quite different. And I am a fool, when I treat a Jew as my equal. I admire and despise them in the same moment. I think, they never should be in a government anywhere, because then we are under foreign rule and we denote through such a choice, that we are not fit and that we need foreign help. The Jew himself in such a position will always be in danger to be a victim of his compensatory power instinct. An uprooted race boils with desirousness, because it has no roots in the maternal soil.[8]

Jung's reply would deeply impact Pitt-Rivers, who clearly interpreted this as an endorsement of his own views. He would repeatedly revisit the letter in his later work and cite Jung as one of his key influences and supporters.

It is thus clear that at this stage Pitt-Rivers' interests still lay in his previous work rather than his present duties. In April 1921, he authored a major report on 'the revolutionary movement in Australia' for his father-in-law, despite the fact that the governor general could do little in the way of policy making. The tone of the finished version was a combination of *The World Significance of the Russian Revolution* and alarmist anti-Communist literature of the period: 'the time has come when the revolutionary incubus must be dispelled or translated into reality. It is no longer a remote peril; it is now upon us, in our midst', it began.[9] The source of this danger was that the 'fresh seeds of internal disintegration were freely and thoughtlessly sown' by Europeans who believed in a 'childishly fallacious misinterpretation' of Charles Darwin's theory of evolution that had suggested 'inevitable and never-ending progress' before the First World War. Now Europe and 'even young Australia' were now on the brink of cataclysmic revolution led by 'subversives' around the world.[10] His proposed remedy for this danger was the suppression of all forms of communist propaganda combined with the creation of a new office headed by 'a trained psychologist having very considerable philosophical knowledge' to 'secure social stability' through censorship, immigration control and legislative advice: effectively a technocrat responsible for suppressing communism through 'scientific' methods.[11]

In the context of the time, Pitt-Rivers' report was only one of a number suggesting that communists had infiltrated Australia's politics at all levels, though it is unlikely many others shared his confidence in the power of psychology to avert revolution. In August 1921, intelligence officials submitted a report warning that communists, mostly foreigners, had infiltrated the Labor Party and were planning imminent revolution.[12] Ironically, only two months later the Party itself had narrowly passed a motion clarifying that it did not believe in the abolition of private property: a direct rebuke to the socialist and communist left that outraged many activists. Australian communism was thus effectively marginalized from the mainstream political parties, though intelligence officials still warned about the potential effects of covert communist propaganda.[13]

In mid-1921, Pitt-Rivers finally found a welcome escape from the stultification and frustration of government work. Leaving Melbourne and his family behind, he sailed for the Bismarck Archipelago north of Papua New Guinea to undertake anthropological field research among the native groups there. While this expedition was unlike anything Pitt-Rivers had done before, it was not a completely unusual decision. His grandfather, after all, had been prominent in the RAI in the late nineteenth century and there were already two anthropological museums bearing the Pitt-Rivers name in England. It is certain that by now he saw anthropological research of his own as the key to a future legacy. As Levy observed openly to his friend, this emerging obsession with anthropology came directly from his adulatory and inflated perception of his grandfather's significance.[14] Privately, Pitt-Rivers admitted that he had another interest in embarking on anthropological fieldwork as well: escaping the confines of his marriage. As he frankly conceded to his father-in-law before leaving the country, his union with Rachel had reached a breaking point and he felt that he needed a period of physical separation from her. This desire was, of course, made significantly more awkward by the fact that Lord Forster was not only his father-in-law but also his superior, and without his permission it would be impossible for him to leave the country. Pitt-Rivers initially suggested a tour of India and Burma but eventually settled on an extended period of field research closer to Australia. Mercifully, his father-in-law approved the leave of absence.[15]

Pitt-Rivers' choice of the Bismarck Archipelago for his research was a wise one, in part because the islands had only recently passed out of German control and become freely available for fieldwork. Specifically, he was interested in two islands that had rarely been examined by previous anthropologists: Wuvulu (also called 'Matty') and Aua (also called 'Durour'). Lying about 150 kilometres northeast of New Guinea, the area had first been explored by Europeans in the sixteenth century but the islands were generally left alone due to their small size and distance from major shipping areas. In 1893, the New Guinea Company, a German venture attempting to establish colonies in the area, dispatched a ship to recruit the natives of Aua and Wuvulu for plantation labour. While this effort was unsuccessful, the expedition leader returned with artefacts from the area that demonstrated a level of cultural distinctness from the mainland and similarities with other small islands to the north.[16] A number of the items subsequently found their way to the British Museum and the Berlin Ethnological Museum, bringing the islands academic attention and the interest of greedy traders.[17] An early attempt to establish a trading post on the island failed when the Danish trader left behind to manage the business was killed by the natives during a dispute. Subsequent German voyagers looted artefacts from the local people with an eye towards resale on the European markets: one collection of more than 3,000 items sold for more than 20,000 Marks in the early 1900s.[18] Suddenly realizing that their material culture could be profitable, the natives of both islands began to openly trade with visiting German collectors and mass produce goods for sale. The result was an outbreak of European-introduced disease that killed off nearly half the population of Wuvulu by 1902.[19]

Similar disaster soon struck the inhabitants of Aua. In 1903, two German traders established a trading post on the island and, when one departed for health reasons the following year, the natives killed the other and threw his body into the sea.[20] Fearing retaliation after the appearance of a European vessel on the horizon, the group responsible for the killing and several nearby villages attempted to leave the island in overcrowded canoes, but a number of crafts overturned and their occupants drowned or were eaten by sharks.[21] Nearly 400 died, reducing the island's population by almost 40 per cent in a single day.[22] The surviving German trader was grudgingly accepted by the surviving natives, eventually marrying the daughter of a community leader and living on the island 'like a native chief' for the next eighteen years.[23] In 1906, a German visitor to Wuvulu commented on the 'disintegration of the island's social fabric' while others remarked that profit-hungry traders had looted both islands of their ethnographic artefacts and the populations there seemed to be experiencing a terminal decline in numbers.[24]

Aua would become the focus of Pitt-Rivers' research in 1921 because, he wrote, 'Until a few years ago her natives had preserved with a greater freedom from European contamination their culture, their language, and their stock.'[25] As the smaller of the two islands – about two miles wide from east to west, for a total of around 3,000 acres, with coral reefs ringing it – Aua had been spared some of the commercial interest that had befallen Wuvulu: 'Up to the time of my visit the Aua islanders knew little of the world beyond the shores of Wuwuloo [sic], they knew no language but their own, and their blood remained almost untouched by foreign admixture, either European or from the surrounding Melanesian groups.'[26]

Intellectually, Pitt-Rivers' initial inspiration in Melanesia was Charles Gabriel Seligman, one of the pioneering anthropologists of the twentieth century. Born in 1873, Seligman had trained as a pathologist and in 1898 volunteered to join an anthropological expedition to the Torres Strait between Australia and New Guinea at his own expense. The Torres Strait Expedition, as it became known, had been arranged by Seligman's close friend Alfred Cort Haddon, who had himself moved from biology to anthropology as the result of a previous trip to the area. Like Haddon, Seligman's interest in anthropology was piqued by his experiences in the Torres Strait and, in 1904, he led his own expedition to British New Guinea.[27]

Seligman's anthropological interests were vast, ranging from psychology to physical anthropology and ethnography, and he published articles in nearly every field he entered. In 1910 he published *The Melanesians of British New Guinea*, an extensive account of his research in the South Pacific, and that same year, he obtained a lectureship in ethnology at the University of London. In 1913, he moved to a permanent position at the London School of Economics, where he became the first professor of ethnology.[28] Haddon remained based at Cambridge, and between those two universities, the men trained many of the leading anthropologists of the twentieth century. Pitt-Rivers had never attended Cambridge or the London School of Economics and possessed little academic training in the field beyond his brief studies at Oxford. He appears to have been given little, if any, formal preparation for field research before going to Aua but he shared Seligman's belief in its importance ('what the blood of martyrs is to the Church', in Seligman's formulation).[29]

Pitt-Rivers' passion and adventurous enthusiasm was no substitute for knowledge, however, and his lack of academic preparation soon showed. One colonial administrator found him 'particularly well read in the theories of Freud and other authorities on psycho analysis [sic]' but ill-prepared for actual research because 'Captain Pitt-Rivers was under the disadvantage, not only of not knowing any native language, but also of being unable to talk or understand our Papuan "pidgin" [a simple dialect of English mixed with words from other languages].'[30] As a result Pitt-Rivers used an interpreter while in the field, leading to at least one humorous but unfortunate misunderstanding. While questioning Papuan natives about their marriage customs through his translator, the male interview subjects misinterpreted his interest in their wives and assured him that they were 'far away, over a distant mountain' when in fact they were in a house in the village.[31]

Some of these criticisms were surely exaggerated. Pitt-Rivers may not have learned native languages in the field, but he did gain at least a rudimentary understanding of several languages and included linguistic aspects in his published research.[32] However, his knowledge of anthropology as an academic discipline – as far as it existed as one at the time – extended little further than his own private reading of a few books on the subject before going into the field. His notebooks from Aua and his other fieldwork do not appear to have survived and were possibly lost in an 'unfortunate accident' that also claimed the first draft of the resulting book manuscript.[33] However, it is possible to gain some understanding of his methods from a number of individual pages of notes from the period and his published work. He appears to have been a diligent documentarian of both cultural artefacts and events he witnessed, attempting to reconstruct historical demographic information about the island from first-hand interviews, records kept by the surviving German settler who had lived out his days on the island, and, later, Australian government statistics. His surviving letters to various government statistical bureaus suggest that he had an almost obsessive interest in numerical measures of population and demography, particularly gender ratios and reproduction rates.[34]

Conditions in the field were harsh. Pitt-Rivers was initially marooned on Aua when the schooner he arrived on was wrecked on the coral reef surrounding the island. A photo from the period shows him sitting outside a small hut in his camp in Southern Papua recording his observations in a notebook, while another shows him dining with travel companions in a room flanked by native servants.[35] His health was precarious: after weeks in the unsanitary conditions of the field, his still-open wound would become inflamed and require medical attention. While he appears to have been able to manage to walk without significant difficulty – photographs show him walking through remote areas without the aid of a stick – extended fieldwork would simply never be possible.[36]

With around six months of field research completed, Pitt-Rivers returned to Australia in late 1921 with anthropology as his new passion and enough 'interesting material' for publication.[37] His sojourns had taken him not only to Aua but also mainland New Guinea and New Zealand,

FIGURE 3.1 *Pitt-Rivers assiduously recorded his observations while conducting his field research. The wound he had suffered in the First World War required constant care and he could only spend a few weeks at a time in the field without medical attention.*

giving him an impressive breadth of experience and knowledge of the groups he had encountered. He also returned with an extensive collection of artefacts, hundreds of which he sent to Henry Balfour, the curator of the Pitt-Rivers Museum, Oxford, who praised the collection as 'very fine' and claimed to immediately make arrangements for its display.[38] It must have been a particular pleasure for Pitt-Rivers to have items of his own discovery placed next to those obtained by his grandfather, and Balfour agreed: 'the fact that the name Pitt-Rivers is associated with the collection is particularly gratifying', he wrote.[39]

The fundamental focus of Pitt-Rivers' emerging research and publications was on the question of why the 'races' he had encountered appeared to be heading towards extinction. He had observed that all over the Pacific – at least the places he had visited – there were unbalanced gender demographics: on Aua, for instance, he claimed that from 1908 to 1921 there was 'a slight, but steady, disproportion in the balances of the sexes in favour of the men, accompanied by a corresponding decline in the total population'.[40] The notion of gender balance, and its implications for a society, would become a major theme of his work, as would his

FIGURE 3.2 *Looking very much the part of a young gentleman, Pitt-Rivers, right, travelled from Australia to his remote research sites by ship. He was joined by Angus Moncrieff for part of the journey.*

interest in the sexual implications of an unbalanced population. Pitt-Rivers was struck by something else as well: in 1921 his former Oxford mentor, McDougall, had proposed a distinction between 'adaptable "extrovert" races' and inadaptable 'introvert races'. Pitt-Rivers extended this distinction to his own work, claiming that South Pacific natives were 'introverts' and that 'all ignorant interference with the social customs and tribal organization of the unadaptable [sic] (these might be the finest and more virile races) would prove fatal to them, and with the advent of our civilisation they tended to disappear'.[41] Evidence of the health of a 'race' could be found in its gender balance. A healthy society would have a surplus of women, he argued, with the practice of polygamy ensuring that all women had male partners and economic support within the community. The fact that European missionaries had abolished polygamy as 'immoral' in many areas, he went on, was at the heart of 'the disappearance...of tribal enterprise and of society in general'.[42] The interference of Europeans, he thus came to believe, was directly driving the Pacific islanders to extinction.

By early 1922, Pitt-Rivers was receiving invitations to present these provocative findings and his extensive collection of photographs to both academic and non-academic audiences around Australia. In March 1922, for instance, he gave a lecture to the Victoria branch of the League of Nations Union under the title 'Native Life in the North-west Pacific, Under the Australian Mandate'. The main thrust of the talk was to argue that he 'is under the opinion that he must try to find out the views of the native on religion, agriculture, and justice before he endeavours to impart our views on these subjects'. He went on to argue that the imposition of European cotton clothing on natives led to 'skin diseases and other ills of the body...and gradually with the advent of civilization the native races die out'.[43] In May, he argued in an interview in the magazine Stead's that Europeans were 'killing the natives with kindness' while 'when they were free to kill one another and to worship idols, their race was preserved'.[44] The best way to save these groups from extinction, he claimed, was 'to leave them alone' and to understand 'the native problem' ultimately required Europeans to 'think black'.[45] In 1923, he presented these findings to the Pan-Pacific Science Congress, where he was listed in press reports as 'an anthropologist of distinction'.[46]

One of Pitt-Rivers' earliest supporters in these academic circles was Flora Marjorie ('Marnie') Masson, the daughter of Melbourne University chemistry professor Sir David Orme Masson. The elder Masson had done specialized research on nitroglycerine and taught at Bristol University before moving to Australia. The entire Masson family was known for its academic achievements: Marnie's brother Irving had followed in their father's footsteps and was a professor of chemistry, eventually becoming vice-chancellor of the University of Sheffield. Marnie herself married a mechanical engineer and became a writer of fiction, history and travel books. Fascinated by the subjects being discussed, in 1922 she was making an effort to attend as many of Pitt-Rivers' lectures as she could.[47] More importantly, in July she sent the précis of one of his lectures to her brother-in-law, whom she was 'sure it will interest'.[48] This was the husband of her sister Elsie, who, in 1919, had married 'a penniless Pole' by the name of Bronislaw Malinowski.[49]

Pitt-Rivers might have heard of Malinowski by the time he set sail for Aua, though it is also possible that he had not. Born in 1884 as an Austro-Hungarian subject, Malinowski initially studied in Krakow and moved to England in 1910 after developing an interest in anthropology and reading the work of W.H.R. Rivers and Haddon.[50] In summer 1910 he registered as a student at the London School of Economics where he soon encountered Seligman, who would become his mentor and academic patron.[51] Their relationship was close, and in 1914 Seligman even helped Malinowski select and pay for the goods he needed for his fieldwork in the Antipodes, including a case of medical supplies containing nearly five thousand pills and other medications for the notorious hypochondriac.[52]

Malinowski's first fieldwork took place primarily in the British-controlled area of Papua, about 1,100 kilometres from where Pitt-Rivers would conduct the bulk of his research a few years later. His main area of interest was the Trobriand Islands, a series of coral atolls with a population of around 8,000 people that Seligman had visited in the course of his earlier research. His timing was poor: with the outbreak of the First World War, Malinowski was classified by the British authorities as an enemy alien and faced possible internment. Narrowly avoiding this fate through the intervention of a friend, he spent the duration of the war conducting research in the Trobriands and other areas alongside extensive visits to Australia.[53] During his visits, he would meet many of the people Pitt-Rivers would later encounter, including the elder Masson and industrial psychologist Elton Mayo, who would become one of Pitt-Rivers' close acquaintances there.[54] Malinowski and Pitt-Rivers had thus travelled in virtually the same social and intellectual circles in Australia at different times. After Malinowski had completed his field research and married Elsie Masson, the pair returned to England in 1920.[55]

The Malinowskis soon moved to the Canary Islands to facilitate the completion of a book, and while there he attempted to write a gruelling 4,000 words a day on the manuscript.[56] By April 1921, the project was complete and appeared the following year under the title *Argonauts of the Western Pacific: An Account of Native Enterprise and Adventure in the Archipelagoes of Melanesian New Guinea*.[57] Examining customs related to gift-giving among tribes in Melanesia, the book was designed, in the words of George W. Stocking, to launch a 'revolution in social anthropology' by encouraging ethnographers to immersively situate themselves in the midst of their subjects to conduct research, 'placing oneself in a situation where one might have a certain type of experience'.[58] The subject of anthropology should not focus on merely cultural artefacts, genealogies and folklore, Malinowski argued, but on how these aspects of societies actually functioned. European observers who did not follow his immersive methods and remained in their own settlements during research, including colonial officials, missionaries and other anthropologists, had simply 'gotten the natives all wrong', he claimed.[59] The role of the ethnographer, in

Malinowski's view, was to put themselves in the mind of their subjects – to understand the world through their eyes as far as it was possible to do so – while maintaining sufficient scholarly distance to return from the field and articulate the lessons learned.[60]

Pitt-Rivers and Malinowski had not yet met in 1921, but it quickly became clear that their work would have commonalities not only because of the geographic proximity of the research sites but also because the two men shared certain aspects of their worldview, including their interest in immersive fieldwork. In late 1922, Pitt-Rivers wrote to Seligman thanking him for the assistance his scholarship had provided ('*The Melanesians of British New Guinea* was one of the few books I took with me into the field') and inquiring whether he would consider supporting his application for fellowship in the RAI, the premier anthropology society in the country and the same organization in which Pitt-Rivers' grandfather had twice been president.[61] Seligman agreed to do so and suggested that Pitt-Rivers should read Malinowski's *Argonauts* for possible affinities with his research.[62] Pitt-Rivers was already aware of the book and had been asked to review it for *The Forum*, an Australian periodical, in part, it seems, because he was acquainted with Marnie Masson.[63] His review was favourable and, in late 1923, Malinowski himself wrote to Pitt-Rivers to thank him for the review and point out the like-mindedness of their scholarly efforts:

> I should like to communicate with you for sometime already, to tell you how extremely stimulating [and] valuable I have found your several utterances sent me in cuttings from my wife's sister. And also to thank you for the very interesting review you wrote (of my 'Argonauts'). I expect I need not add how very congenial your outlook and your general Weltanschauung [world view] is to me and that at a time when anthropology has become a stony desert.[64]

Malinowski went on to ask Pitt-Rivers for assistance with the training and preparation of a colleague who was seeking to do field research in the South Pacific. Despite the man also being a Roman Catholic priest, Malinowski wrote, he 'understands well, that he would have to leave all R.C. nonsense behind and especially not to antagonize white or black, yellow or brown'.[65]

These statements highlight an important similarity in Malinowski and Pitt-Rivers' scholarship and outlook: hostility to the establishment and religion. Both men despised the missionaries they encountered in the field, believing them to be doing more harm than good in many cases and holding the type of data they had produced for an earlier generation of ethnographers in contempt.[66] Pitt-Rivers had an abiding dislike for Christianity as a whole, which he intellectually justified through his anti-Semitic reading of Nietzsche, while Malinowski took a characteristically functionalist view of its role in society and had himself closely read Nietzsche in his younger days.[67] Levy described Pitt-Rivers as a 'wolf' among the sheep of Christian missionaries in the Pacific, assuring him that, 'no better Nietzschean missionary could have been sent...for if I am not mistaken you will seriously disturb the peace of other missionaries'.[68]

In a similar sense, Malinowski was a sort of rebel figure in the anthropological establishment of the 1920s. Many older leaders in the field still believed in classic evolutionist theories – not unlike the views held by General Pitt-Rivers – and the functionalist model Malinowski and Pitt-Rivers espoused was still very much a minority position, if it could be properly classed as a full-fledged theoretical position at all.[69] In essence, the emergent functionalist view argued that all aspects of human society – including taboos, social rules, religion and family units/kinship – were created by societies to fulfil basic human needs. Rather than studying the evolution of these social institutions over time, as the earlier model had suggested, functionalists believed in examining how these aspects of society helped provide for its needs. If a given institution successfully provided an advantage to the society, it would be reinforced and become permanent.[70] This was not to say that a taboo or custom was 'right' in any spiritual or moral sense but merely that it served a function within that society.

This intrinsic iconoclasm towards established interests and ideas appealed to the rambunctious Pitt-Rivers, particularly regarding religion and family structures. He was a life-long critic of the church, and functionalism provided an avenue for him to academically criticize religious ideas by attacking their social outcomes, as he would later do through his involvement in the eugenics movement. Secondarily, his interpretation of functionalism suggested that there was no absolute 'right' or 'wrong' family structure or marital relationship. In his view, Western ideas of monogamy were not intrinsically correct but had emerged because of the social benefits they once provided. As will be seen, in Pitt-Rivers' view this meant that a new conception of the family that dispensed with traditional notions of monogamy and sexual morality might now arguably offer greater social advantages.

In addition to their intellectual affinities for one another, there would be no doubt about the strong personal relationship between Malinowski and Pitt-Rivers. Both men possessed a certain social snobbery that they took into their academic circles. Malinowski once admitted to his future wife that he 'would look up a man in Who's Who to see whether his letters are A.1 at Snob's Lloyds & whether his [Oxbridge] College is incorporated or affiliated'.[71] The fact that Pitt-Rivers had attended Eton and was the heir to one of the largest estates in the country must have held an appeal to his new friend. Malinowski would soon be responsible for introducing Pitt-Rivers to the key academic circles developing anthropological functionalism, giving him a seat at several significant tables. At the same time, it was a tempestuous relationship, in part because of the men's differing politics. Malinowski's inclinations were towards the left and 'champagne socialism' while Pitt-Rivers was conservative, militantly anti-communist and increasingly moving towards fascism as the 1920s progressed.[72]

In the meantime, however, Pitt-Rivers had established himself as not only an anthropologist but also as a demographic researcher due in large part to his interest in statistics related to gender and birth rates among the natives on Aua and elsewhere in the Pacific. He joined the prominent Australasian Society for the Advancement of Science and was quickly appointed to leadership roles in several commissions including the Committee on Vital Statistics of Primitive Races.[73] Throughout the mid-1920s, he sought the most updated demographic information on the Maori population and other native groups, clearly intending to pursue a wider research project than his field research had suggested. The demographic numbers he collected predictably showed that the ratio of natives to Europeans was moving in a direction unfavourable to the former.[74]

It is clear, however, that Pitt-Rivers had always desired to return to England as quickly as possible. His posting in the country had always been seen as temporary and for the duration of his father-in-law's duties at the longest. The fact that he had asked Seligman to support his application to the RAI was evidence that he anticipated an imminent return to England, as was the fact that he published his first significant article on his South Pacific research in the RAI's in-house journal in 1925.[75] As it turned out, he did not have to wait long before an opportunity to return home presented itself. With his father-in-law's term as governor general coming to a close, Pitt-Rivers departed Australia in January 1925, arriving in England by May after a four-month journey that included stops in the Pacific and the United States.[76] Upon returning, he immediately began busying himself with the RAI, where he volunteered to read a paper at the earliest possible opportunity and provide materials for rapid publication.[77] Invitations to speak to the Cambridge Anthropological Club were forthcoming from A.C. Haddon, as were invitations to meet and talk from Malinowski.[78]

Despite his increasing academic profile, at this stage Pitt-Rivers still lacked the academic qualifications that would be required for an academic post. Formally speaking, he had only his Eton education when he returned to England in 1925, with his military service, brief time at Oxford that had resulted in no degree, and work in Australia making up the bulk of his *curriculum vitae*. The regard he enjoyed among Haddon, Seligman and Malinowski was derived in part from his family name, personal connections in Australia and the reputation of the Pitt-Rivers Museum in Oxford. His academic publications to date were impressive but he lacked the work that would make him an academic equal of the anthropologists he revered. He intended to rectify this situation, and the period until 1930 was incredibly productive for him. From 1925 to 1927, he again based himself in Worcester College, Oxford, immersing himself in the heady atmosphere of the Senior Common Room. In addition to speaking engagements and academic lectures, he published a number of academic articles and interventions derived from his field research, mostly in the RAI's journal *Man*.[79]

At a RAI meeting in 1926, Pitt-Rivers delivered a paper on the 'effect on native races of contact with European civilization', no doubt rehashing his now well-practiced arguments. In the audience was Arthur Keith, one of the most famous scientists of the age. Already in his early sixties, Keith was a living legend in the British anthropological community, had served as president of the RAI in the 1910s and had known Pitt-Rivers' grandfather well. He had started his career as an anatomist, discovering the pace-making sinoatrial node of the human heart and later became interested in human evolution, studies of the human skull (craniometry) and palaeontology. In 1908, he was elected as conservator to the Royal College of Surgeons, which put him in charge of its museum collections. His primary interest was collecting skulls from different races around the world to compare their sizes and qualities.[80]

Keith was convinced that modern human races had separated from one another early in the evolutionary process and were by nature in competition with one another for resources. Culture and prejudice, he believed, were evolutionary mechanisms that discouraged competing groups from interbreeding and allowed them to maintain their racial distinctiveness. Much of his research was directed at uncovering the origins of human races and in 1912 he was on the scientific team that examined the 'Piltdown Skull', a set of skull fragments found in Sussex that had been assembled into a partial cranium. Keith was among the first experts to see the reconstructed skull, which appeared to have the lower jaw of an ape and a brain capacity similar to that of modern humans. It was an immediate scientific sensation: 'Piltdown Man' was hailed as a 'missing link' between ape and man, the 'the Earliest Englishman', and the first representative of an ancient race in the British Isles.[81] However, because the skull was fragmentary, there were multiple ways to assemble its pieces. Keith argued that the initial reconstruction had been flawed and his own subsequent reconstruction made the brain cavity even larger and roughly on par with modern *Homo sapiens*.[82] Sceptics argued that both reconstructions were inaccurate, and some suggested that the 'skull' was nothing more than an ape mandible combined with a human cranium.[83] They would be proven correct in 1953 when the Piltdown Skull was exposed as a deliberate and sophisticated forgery committed by a still-unknown individual.[84]

However, the idea that a putative Piltdown Man to whom the skull had belonged was an extremely ancient missing link between apes and modern humans fit with Keith's overall view of human racial divergence: if the fossil was indeed from an early evolutionary forerunner of modern Britons, it would have supported his view that modern races had diverged from one another early in the evolutionary process and developed mostly separately.

He argued that racial isolation was the key to evolution, and racial prejudice was nature's way of maintaining the group's biological strength in the face of 'degeneration' from outside interbreeding. In 1930, he delivered an address to students at Aberdeen University describing the world's races – 'white, yellow, brown, and black' – as equivalent to the uniforms of competing football teams ('No transfers for her [nature]; each member of the team has to be home-born and home-bred... she made certain that no player could leave his team without being recognized as a deserter.').[85] Racism was the way nature motivated these 'teams' to compete against one another for resources, he claimed. 'What modern football team could face the goal-posts unless it developed as it took the field a spirit of antagonism towards the players wearing opposing colours? Nature endowed her tribal teams with this spirit of antagonism for her own purposes', he told the students.[86] In the assessment of historian Elazar Barkan, Keith was 'the most prominent racist among British anthropologists' in the first half of the twentieth century.[87] The fact that he would become one of Pitt-Rivers' key supporters in the scientific community of the day would be an essential element of the latter's success.

In this context, it is hardly surprising that Keith was interested by Pitt-Rivers' paper on racial decline and extinction. After all, his view of racial competition entailed that some races should eventually lose out entirely in the evolutionary process. He was most impressed, however, by the intelligence and mannerisms of the young man making the presentation and saw echoes of his famous grandfather. Pitt-Rivers, he later wrote, 'brushed sentiment aside and laid bare in all their nakedness the causes which he found at work amongst the decaying races of the Pacific'.[88] He continued:

I was struck by his precision of method and of statement, by his insistence on calling things by their proper name, his hatred of Euphemisms, his impatience with those in authority and his inclination to force the truth home by hard knocks rather than by gentle persuasion. Clearly a young autocrat – with nothing of the democrat in his composition...

I was not surprised to find that my young friend, Captain Pitt-Rivers,... was the grandson of the great General; both carried themselves and their clothes with an air of distinction. Both approached their selected problems with a military directness. Both had the same impatience with stupidity and slack-thinking.[89]

This was evidently an insightful description, as it was later quoted by one of Pitt-Rivers' paramours as necessary 'to obtain any idea of the kind of person Joe [sic] really was' because he was 'never given to suffer fools gladly. He was far too blunt for that'.[90] Impressed by both the intellect and the mannerisms of his new friend, Keith would become a lifelong friend and supporter of Pitt-Rivers' work.

The following year, Pitt-Rivers completed and published the book that would effectively amount to his *magnum opus*. Based on the papers he had been presenting for the previous few years, the work was an in-depth account of his experiences and research findings in the Pacific, carrying the extensive title *The Clash of Culture and the Contact of Races: An Anthropological and Psychological Study of the Laws of Racial Adaptability, with Special Reference to the Depopulation of the Pacific and the Government of Subject Races.*[91] Pitt-Rivers had been working on the book since 1924, when he was still languishing in Australia, and by the time he left for England three sections examining 'race extinction', 'sex-ratio variances' and 'culture extinction' had been completed. However, the entire first manuscript was lost when Pitt-Rivers left it on a railway platform.[92] Starting again from scratch, he must have realized that this organizational structure was insufficient and made significant changes to the scope of the study. The first half of the completed work dealing with 'the native problem', the 'decline of subject races', 'the influence of miscegenation' and polygamy were deemed sufficient for the awarding of a research Bachelor of Science (BSc) degree from Oxford following the completion of revisions suggested by ethnologist R.R. Marett, while additional sections were later added for publication.[93]

The finished book had three main audiences: academic anthropologists, imperial administrators dealing with 'subject races' and the reading public. It was not merely a work of ivory tower academia; Pitt-Rivers intended it to be a policy prescription as well. Stocking has described *The Clash of Culture* as 'a melange of the intellectual tendencies available to anthropology in the early 1920s, including Freudian and Jungian psychoanalysis, racial ethnology, and eugenicism, as well as functionalist anthropology'.[94] Indeed, the book's frequent mentions of Nietzsche and Freud were characteristic for Pitt-Rivers. The final chapter, entitled 'The White Man's Task', contained a scathing attack on Christian missionaries and administrators he blamed for the cultural breakdown he had seen in the Pacific. 'Christian proselytism has done irreversible harm to native races by disintegrating their culture, and to us also by unrest and antagonism the process evokes', he argued.[95]

Identifying himself with Malinowski's emerging functionalist school (and dedicating the book itself to Malinowski himself), Pitt-Rivers argued that the culture level of a community was conditioned by three factors: culture-forms (traditions, beliefs, customs, art), culture-accessories (weapons, tools, artefacts more generally) and culture-potential (innate abilities to develop new innovations).[96] He believed this latter category was most important because it dictated how a given culture *might* develop in the future, as opposed to what and how it had already developed. This notion, of course, was closely related to his previously expressed view on the role of Jews in the Russian Revolution: even if they were not consciously aware of it, every Jew was responsible because of their shared mental inheritance. In the same way, a racial group's innate 'culture-potential' dictated all of its future development.

The implication of this, he claimed, was that indigenous groups with little 'culture-potential' would inevitably be destroyed by contact with Europeans because there was no possible way to 'raise a people in cultural level' beyond their innate potentiality.[97] The interference of outsiders trying to Europeanize these groups, particularly missionaries, he claimed, generally resulted in the extinction of the lesser groups as they lost the will to survive and were separated from their customs. The tell-tale sign of a group's health could be found in its gender ratios: a surfeit of males suggested that a group was in terminal decline because there would be increased competition for mating partners, while a surplus of females suggested a healthy group where reproduction rates could be high. Every tribe he had observed suffered from an excess of males, leading to an unbalanced breeding situation. Sexual mores were at the centre of this: by forbidding polygyny (the partnering of multiple women to a man) out of moral concerns, European administrators had disrupted the 'marriage system' of these groups and harmed their ability to maintain their numbers.[98] In contrast, Pitt-Rivers believed that polyandry (the partnering of multiple men to a woman), which might be seen as the obvious solution for a society with a surplus of men, was counterproductive because female overexposure to sperm 'is unfavourable to, or even inhibitive to, fertilization... the well-known small tendency of prostitutes to become pregnant, which cannot be altogether attributed to chronic venereal disease, would also be accounted for by an immunization through over-exposure... it may be supposed that the spermatozoa of different individuals have a counteracting effect on one another'.[99] In his reckoning, societies with an excess of males therefore had no biological way to counter their decline, while groups with an excess of females could easily increase their numbers.

Pitt-Rivers' conclusion was that European interference in native societies was almost entirely responsible for their rapid decline following contact. Fundamentally, native customs should be left alone when possible: 'It is quite clear that we do not need to destroy native customs, even though they may appear unpalatable, that is if it can be shown that these customs are indispensible for the integrity of native culture.'[100] The idea that natives could be 'raised to the level' of Europeans was ludicrous, he argued, because of the differential culture-potential in each group. 'We are loath to acknowledge that we cannot "raise people to our own high cultural level" by changing their culture-forms. Yet facts prove that culture-potential cannot be modified without first modifying blood, though that of course follows when races are in contact', he wrote.[101]

He concluded the study with a provocative call to action:

Is it too late to hope that now by studying more sympathetically and intelligently native customs and ideas, we may learn their intrinsic value as expressions of social purpose? To the rising generation of our dark-skinned subjects have we the right to say more than this? May they learn to value whatever is sound or beautiful in their own culture, in which may be found the surest promise of their own racial achievement in place of blindly following the lead of people whose proffered cultural gifts they can never truly make their own.[102]

This was a strong critique. Pitt-Rivers was calling for the cultures and practices of these groups to be effectively protected from 'interference' by outsiders that would inevitably result in their destruction. As Stocking has observed, Malinowski had leanings in this direction as well: in 1916 he had presented a report to an Australian parliamentary committee in which he argued that natives should be left alone, 'to their own conditions', for their own good. Malinowski's view was that these populations should be subjected to 'scientific management' for their protection.[103]

In these contexts, there is a temptation to lionize Pitt-Rivers' arguments for the protection of native practices. To an extent, at least, he was a cultural relativist, believing that customs should be evaluated within their own contexts rather than through the lens of another group's values, and he clearly agreed with Malinowski that European interference in native practices was exclusively destructive.[104] Right-wing artist and writer Wyndham Lewis later described his book by writing that, 'Politically it represents, surely, propaganda for tolerance. Would that the Anglo-Saxon could sweat out his moral virus. But it would be too late, if he did, to save all the "brave little" black, brown, and red nations he wiped out.'[105]

On the other hand, however, it is critical to remember that this call for cultural tolerance was fundamentally born from the view that 'races' were not biologically equal. Pitt-Rivers' very idea of 'culture-potential' was infused with a racist set of assumptions about what groups would be capable of psychologically achieving in the future. While the preservation of native practices and cultures might be seen as an admirable outcome of this view, for Pitt-Rivers these efforts were based on the recognition that these groups were irreparably inferior. This idea of 'separate development' for races might even be seen as having similarities to the *apartheid* policies pursued by the South African government after 1948 in which non-whites were forced to become citizens of 'Bantustans' and expected to 'develop separately' from whites. Further, while saying relatively little about the Jews, Pitt-Rivers made a point of explicitly stating that his theories applied to them as well, claiming that their passionate 'temperaments' predisposed them to diabetes, a 'Jewish' disease.[106]

Indeed, for these reasons Pitt-Rivers expressed deep concern about the potential effects of racial mixing. Some races were simply 'incompatible' with one another and could not profitably mix, he argued, and 'more powerful culture bearers' would inevitably wipe out 'less powerful' cultures.[107] At the same time, however, the more powerful races would become assimilated or polluted by breeding with lesser races.[108] This danger was biological as well: mixing, he claimed, directly affected the sex ratio of the next generation and might even lead to the extinction of the hybrid population as it became dangerously low in a gender or less adaptable to its environment as a result of obtaining new hereditary qualities.[109]

Predictably, *The Clash of Culture* set off an immediate firestorm in both academic and government circles. In the months following its appearance, Pitt-Rivers received numerous letters from imperial administrators praising his insights and thanking him for arguing points that they felt had been underrepresented. One such correspondent stated that *The Clash of Culture* 'should be used as a handbook to Administrators' all around the Empire.[110] An appraisal in the prestigious *Journal of Heredity* praised it as a 'well-documented critical review' of the reasons for native decline in the Pacific and supported his recommendations for improving British imperial policy. 'Before attempting to administer the affairs of a "subject race", study thoroughly the native culture', the reviewer concluded from the work. 'The science of anthropology can help here. Secondly, minimize the contact and destroy as little of the native culture as possible.'[111] Malinowski and his colleagues were also impressed. He liked Pitt-Rivers personally and admired his field work and writing, mentioning him to his wife as one of his 'male admirers', though he also, no doubt accurately, described him as having 'the foibles of a spoilt child'.[112] In 1927 he used Pitt-Rivers as an external examiner for his star student, Raymond Firth's, doctoral thesis.[113] Firth was later to become one of the leading anthropologists of the century and sensibly decided to court his examiner by favourably citing *The Clash of Culture* in his thesis and the resulting publication.[114]

Yet this adulation was hardly universal: renowned zoologist and eugenicist Julian Huxley, to whom Pitt-Rivers had sent drafts of the work for review, argued that the heart of his research on sex ratios as an indicator of racial health had 'insufficient biological foundation', thus effectively calling into question the premises upon which the entire argument had been built.[115] Several colonial administrators Pitt-Rivers had encountered during his field research were outraged by what they perceived to be a betrayal of their trust and believed that he was deliberately attacking them. The Minister for Home and Territories in Canberra immediately sent a letter to the Lieutenant Governor of Papua, Hubert Murray, asking for clarification of Pitt-Rivers' claims, specifically those related to the government's stance towards polygamy. Pitt-Rivers had stayed at Murray's residence during his research, in part because 'he was, in a sense, under the aegis of the Governor General' but evidently never interviewed his host to clarify factual questions and had shown no knowledge of local languages.[116] Murray personally disliked him and defensively argued that his understanding of colonial policy and the natives themselves was a result of his personal biases rather than facts:

Among his prejudices was a strong bias against all forms of Christianity and all forms of missionary enterprise, and still a stronger dislike to anything Australian and especially to Australian democracy ... his method seemed rather to grasp any chance rumour that seemed to support his prejudices, and to ignore all evidence to disprove them.[117]

Murray went on to cite other provocative statements Pitt-Rivers had made after returning to England, namely his claim that in Europe unmarried women should be free to 'receive lovers' in an 'apprenticeship for marriage', as was practiced among some South Pacific groups. While this argument was in fact more nuanced and was actually arguing for the state to recognize unions outside of traditional marriages, it was reported in *The Sunday Times* (Sydney) under the sensationalist headline 'Tribal marriages! Captain Pitt Rivers' Amazing Sexual Sermon – What the girls do in Melanesia'.[118] Sending a clipping of the article to his superiors, Murray concluded that 'I think I have given ample evidence that Captain Pitt Rivers' statements can not be taken seriously.'[119]

Regardless of whether Murray was correct about personal prejudices dominating his research, it is true that a distinct political agenda was never far from Pitt-Rivers' thoughts and writings. Malinowski was frequently annoyed by his tendency to constantly deviate conversations to discussions about Nietzsche, religion or political matters, telling his wife in late 1927 that a dinner with Pitt-Rivers in Oxford had culminated with a discussion of his 'political plans and ambitions'.[120] Just days earlier, the men had clashed in a heated debate over religion, as Malinowski recounted to his wife:

We have now had a long standing argument about Christianity in which I maintain that the really aristocratic principle is to keep God well measured for the lower classes and to keep agnosticism for the select few. Therefore Christianity especially Roman Catholicism should not be combated, but rather new religions and pseudo-religions. He doesn't see it and we argue, as we did before, about a certain political sect – (the 'Bloody Fools'). But I feel at times the futility of it all, when I realize that Jo is one of my best friends here and how very far he is from me.[121]

The 'Bloody Fools' were the British Fascists, the country's first self-proclaimed fascist party.[122] Indeed, by 1927 Pitt-Rivers' politics had already begun to shift even further to the right. His friendship with Oscar Levy had played a role in the change: following Italian dictator Benito Mussolini's 1923 'March on Rome' and the establishment of a fascist government in the country, the men had discussed whether they had finally seen the advent of the world's first Nietzschean government. 'Mussolini is indeed the most interesting of all European figures', Levy told Pitt-Rivers in 1924. 'I have made some inquiries about him and am now sure that he is a disciple of Nietzsche .... We Nietzscheans ought to get in touch with him. We are his spiritual brethren of whom he himself stands in need.'[123]

Three months later, Levy breathlessly wrote to Pitt-Rivers reporting that he had met Mussolini in person and confirmed the *Duce*'s affinity for Nietzsche:

First of all – what I always suspected, has been confirmed. Mussolini's ideas are largely drawn from Nietzsche. I had written him a long letter, telling him about the Nietzschean 'mode' in all he said and wrote – and he began by telling me that he had read all his works fifteen years ago and that he was deeply impressed by them .... He spoke to me about twenty minutes, absolutely frank and without any witnesses, about Nietzsche, Machiavelli, human nature, universal suffrage .... He told me that he had quoted in his previous writing both Nietzsche and Machiavelli, but that he should do so no more, as he had had no end of misunderstanding from that. 'People are too stupid.' 'They are also downright bad: Machiavelli knew men well.'[124]

An account of Levy's interview with Mussolini was later published, with the *Duce*'s approval, in *The New York Times,* and concluded with the line 'Let us learn from Italy and Mussolini!'[125]

Pitt-Rivers was similarly convinced that Mussolini offered a potential solution to the problems he had diagnosed in European politics. Claiming that fascism represented an end to 'class dictatorships' including democracy, he evoked Nietzschean language to describe its innovations:

The first thing, then, to recognize about the Fascist revolution in Italy is that it is a revolution in *ideas* of government, a transvaluation of values. It creates a new departure in ideas of government. This does not mean that the ideas at the back of Fascism are new to the world. They are certainly not that, since the principles of good government, of statecraft as a science, have been known and written about since the dawn of the written history of our civilization, but they are new, only in the sense that they form a definite break with a comparatively recent departure in the tradition of the institutions of the government. This recent tradition, not being based upon any first principles, has been fluid, neither clearly recognizing where it came from, nor where it is leading to, nor upon what it is based. Vaguely we talk of this fluid and everchanging conception of political governance as 'Democracy'. Actually this vague and fluid conception of democratic parliamentary government has led to the formation of governments by political groups banded together in order to force their particular class or group interests upon the whole of the nation, and alternatively calling themselves Conservative, Liberal, Socialist, or Communist.[126]

The result of fascism's triumph in Italy, he continued, was that 'the idea of efficiency in government has been made a principle' and 'the idea of functional government and functional representation is restored'. The greatest danger to fascism's success lay in Mussolini's tolerance of the Catholic Church and the fact that he 'courts the Vatican on every occasion'.[127]

Despite his past troubles with the Home Office, Levy was granted a visa and readmitted to England in 1925 following a direct intervention by Pitt-Rivers on his behalf.[128] Both men were thrilled with the outcome and began discussing their next collaborative effort, though they would never again publish together.[129] As shown by his increasing interest in Mussolini and early writings on fascism, by 1928, Pitt-Rivers was already on the road departing the mainstream scientific establishment as his views became more extreme. Levy, for his part, would soon express regret for his own interest in Mussolini and the effect it had exerted on his friend.

CHAPTER FOUR

Eugenics and the Science of Population

Though he could not have known it, by 1929 George Pitt-Rivers'
academic career had nearly reached its pinnacle. He was now a fellow
of the RAI and had been invited to both present papers and serve as a
respondent to other presentations. Being elected to the presidency of the
organization, which was not an outlandish ambition at this stage, would
have put him on par with the achievements of the grandfather in whose
shadow he was working. However, there was still more groundwork to
be done first. While *The Clash of Culture* had been sufficient for a BSc
degree from Oxford, it was seen as insufficient for the university awarding
him a Doctor of Philosophy (DPhil) without further academic work.
There was now an impetus for him to produce as much as possible to
obtain the doctorate and presumably move on to an academic post from
there.[1] He had ample personal support for his ambitions and was a close
friend of Bronislaw Malinowski despite their political disagreements.
In early 1929, for instance, Malinowski took a break from his rigorous
work schedule to drink with Pitt-Rivers and 'two Yanks' until the early
hours of the morning, ruining his work ethic for the following day but
evidently still pleasing him.[2]

Later that year, Pitt-Rivers made an extended visit to the United
States and relied on his academic friends for introductions into American
university circles. Writing to prominent University of California, Berkeley
anthropologist Robert H. Lowie, Malinowski introduced his friend as an
intelligent and distinguished representative of the British establishment who
nonetheless harboured some eccentric views:

A few weeks after this letter arrives, you will probably have a visit from a
great personal friend of mine, Captain George Lane-Fox-Pitt-Rivers – the

only man in England privileged to possess a name with three hyphens, a Fascist Weltanschauung, my friendship, considerable intellectual and personal charm, and three or four Dukes in the past few generations of his lineage. In spite of all these draw-backs I think you will like him and appreciate his mind and personality.[3]

Malinowski's introduction to Lowie had the desired effect and in mid-1929, Pitt-Rivers gained an introduction to the Berkeley anthropology department. Lowie himself had read *The Clash of Culture* and was impressed by its conclusions, seeing Pitt-Rivers as a kindred spirit and doubtless extending great generosity in California to his guest.[4] It did not go unnoticed: 'I must... briefly place on record my very special and grateful thanks for all the kindnesses you showed your temporary wandering adoptive sib-member', Pitt-Rivers told Lowie following his visit to California. 'Long shall I cherish the recollection of them.'[5] Despite the hospitality, it is clear that Pitt-Rivers was cultivating a reputation for not only his intellect and charming personal manner but also his extreme political views. As long as these remained secondary to his academic work they could be overlooked; Malinowski, as already noted, was hardly a creature of the far right and saw politics as one of the main areas where he and Pitt-Rivers did not agree but for now they remained close friends regardless. The eventual break between them would bring their political differences to the fore.

In the meantime, however, *The Clash of Culture* had been generally well received and garnered continued praise. Malinowski cited the work in an article on 'Marriage' in the *Encyclopaedia Britannica*, praising it as a 'most important contribution to this subject', while C.G. Seligman cited it as a ground-breaking study on the New Zealand Maori in lectures at The Royal Institution in London. As late as the 1950s, anthropologist and Malinowski protégé E.E. Evans-Pritchard wrote that the study had 'long been recognised as an original and outstanding study of primitive societies'.[6] Pitt-Rivers had achieved the academic recognition he had long craved, and by the late 1920s, he could reasonably expect that his name would eventually sit next to his grandfather's in the annals of scientific history. The rhetorical excesses of *The World Significance of the Russian Revolution* could now be forgotten amidst his growing publication list. Though anthropology was still an emerging field and seen as academically suspect by many, Malinowski, Seligman, Lowie or their colleagues could presumably have been counted on to lobby for the academic hiring of such a distinguished young talent: he had, after all, already served as Raymond Firth's doctoral thesis examiner.

In a dramatic show of support, Arthur Keith purportedly told John Linton Myres, Chair of the Committee for Anthropology at Oxford, that social anthropology should immediately be given an Honours School with a full chair and that Pitt-Rivers should be recruited to fill it.[7] This would have been an almost incredible step given that the university had only established the subject as a supplementary topic in 1885 and began offering a diploma in the area a full twenty years later. Myres was evidently outraged by this attempted outside interference in the university's affairs and no action was taken.[8] All Souls College eventually created the university's first Professor of Social Anthropology in 1936 and the university appointed another South Pacific anthropologist and avowed functionalist, Alfred R. Radcliffe-Brown, to the post.[9] Pitt-Rivers never forgave the insult and later accused Radcliffe-Brown of plagiarizing his work.[10]

In the meantime, events had already begun to conspire against Pitt-Rivers' academic career. In August 1927 his father, Alexander Pitt-Rivers, died, leaving him 'the responsibility of owning and caring for the interests of cultivators of 20,000 acres of the fairest agricultural land in England'.[11] Though not unexpected, his father's death just an hour after Pitt-Rivers had arrived at his bedside sent him into a bout of philosophical self-doubt. He confided his feelings to Malinowski in a moving letter expressing unusual personal insecurity:

It was curious that for some days before I had been oppressed by thoughts of failure, and the futility of the ridiculous little span of life during which we busy ourselves frantically in efforts, which, even if they have any durable effects at all, can, in their insignificant totality, but have some tiny indelible mark – a mere scratch – on the vast rock-monument of time. My father's death shocked me more than perhaps I had anticipated. It was, I supposed, less the sudden severance with an intimate and life-long familiar figure, for my father, kind and warm-hearted though he was to me as a child and as I now look back on him, never knew me intimately as a man, he seldom knew where I was, had but the vaguest idea of what I was doing, was incapable of talking or thinking in the same language, and could even less find or take any interest in a world of thought or action so remote from his own. But the fact that his whole life... had been solitary, isolated, uneventful, monotonous and without achievement, made its culmination in slow-creeping death, after months and years of flickering, partly-conscious, dependent helpless life, seems to me an epitome of tragedy and fruitlessness. A potentiality for achievement unrealized – yet when it is realized does it not end in the same awful obliteration and silence?[12]

This remarkable honesty could only have been shared with a truly close confidant, and there are few surviving letters from Pitt-Rivers expressing similar sentiments and vulnerability. These were the statements and questions not only of a man in the throes of mourning but struggling with deep existential questions.

Unlike his father, Pitt-Rivers intended to reside at the family's manor house at Hinton St Mary rather than at the Rushmore estate of his grandfather. The Hinton house was in some ways grander but in need of repairs, and he soon embarked on an expensive project to restore it. The renovation was time consuming but created a perfect venue for hosting and entertaining guests that he would soon utilize to maximum benefit. His family's fortunes and future now lay almost completely with him – not to mention the welfare of his many tenets – and there was no chance of forsaking these responsibilities. The unfortunate corollary was that the academic world would now have to take a backseat to these obligations.

To add to the complexity, after 1929 Pitt-Rivers' personal life took several significant turns. He and Rachel had now been married for nearly fourteen years and had two sons, though he had openly admitted to his father-in-law nearly a decade earlier that their marriage was in trouble. The fact that he had left his young family for months at a time to pursue field research had presumably not helped the situation. Beyond attending to her household and her official duties as the governor general's daughter while they were in Australia, 'Ray' had increasingly developed a passion for acting and had taken to the stage when her husband left for his research. In 1929, his restorations of the Hinton St Mary estate included the construction of a small theatre in the ancient Tithe Barn on the property for both her use and local community events.[13] Under the stage name 'Mary Hinton', Ray had already appeared in a number of productions and several films.[14] By the end of 1929, however, their marriage had completely collapsed and in April she filed for dissolution.[15] Pitt-Rivers did not contest the suit and in January 1930 their marriage was officially ended, with Rachel obtaining legal custody of their children.[16] In the years to follow, she would go on to a film and television career that included more than three dozen credited screen appearances through the 1960s.[17] Their relationship following the divorce was strained but would have to be maintained in part because Michael, their eldest son, was Pitt-Rivers' presumptive heir.

Pitt-Rivers soon married for the second time, and his choice of a new wife would in part be a testimony to his own scientific pretensions. In 1931, he wed Rosalind Venetia Henley, a promising biochemist who, unusually for a woman at the time, had recently completed both bachelor and master of science degrees at Bedford College, University of London. Born in 1907, 'Ros', as she was often called, was nearly two decades younger than her husband and was descended from two prominent English families: her father, Anthony Henley, was a brigadier general in the First World War before dying at age 52, allegedly during a cricket match in Bucharest.[18] More significantly, her mother was descended from the Stanleys, the same family into which General Pitt-Rivers had married. Ros' grandfather had been 4th Baron Stanley of Alderley and her mother, Sylvia, had 169 first and second cousins, many of whom were prominent. One of Sylvia's aunts, Alice, had married General Pitt-Rivers in 1853 and was therefore George Pitt-Rivers' grandmother, making Ros distantly related to her husband. As already noted, another aunt, Katherine, became Viscountess Amberley and was the mother of Cambridge philosopher Bertrand Russell. A third aunt, Blanche, became Countess of Airlie and Clementine Churchill's grandmother. Sylvia Stanley's younger sister, Venetia, was known for her beauty and became the platonic confidant of Prime Minister Herbert Asquith during the First World War, with him eventually writing to her four times a day about governmental and military matters. Though she later married another man, on her death it was discovered that she had carefully preserved more than 500 of Asquith's letters and substantially fewer from her husband or her other lovers, who included Canadian newspaper magnate Lord Beaverbrook.[19]

These family connections also meant that Ros, like her husband, was related to Baron Redesdale and the Mitford family, several members of which would become well-known in the 1930s for their extremist political views. One of Lord Redesdale's daughters, Diana Mitford, would become British fascist leader Oswald Mosley's second wife while another, Unity, became an outspoken Nazi sympathizer and a close friend of Adolf Hitler.[20] It is perhaps unsurprising that given these extensive and prominent family ties, Ros spoke often of her Stanley forebears and far less often about her father's comparatively modest lineage.[21]

In August 1932, Ros gave birth to a son, George Anthony, and temporarily abandoned her plans to pursue a doctorate and further research. She moved to the Hinton St Mary manor house to begin attending to domestic matters and would not resume her scientific work until after the breakdown of her marriage. Indeed, from the first days there was tension between her family and her new husband: Ros' mother, Sylvia, quickly took a disliking to her son-in-law and he vocally reciprocated the animosity.[22] The fact that Ros had a great deal more scientific potential and was already better academically qualified must have also frustrated her husband. Certainly their post-war careers would leave little doubt as to which of them had a greater scientific impact, with Ros being elected to the Royal Society in 1954 in recognition of her ground-breaking work on thyroid.[23] In the meantime, however, it was her husband who enjoyed academic status and distinction. With his reputation as an anthropologist now established, Pitt-Rivers increasingly believed he had the knowledge to apply his South Pacific research to social questions in Britain. He soon embarked on an effort to extend his anthropological theories to contemporary society. It was through this effort to make his views of racial decline and extinction relevant to modern Britain that he became involved with the eugenics movement.

The British eugenics movement was a strange hodgepodge of ideas and competing social agendas by the early 1930s. Based in part on the nineteenth-century work of Francis Galton, a cousin of Charles Darwin, eugenics was originally predicated on the idea that the inheritance of traits between human generations could not only be predicted along mathematical lines but should also be controlled to maximize favourable traits, similar to practices in animal husbandry. For Galton and the early eugenicists who embraced his ideas, traits including intelligence, athletic prowess, memory and virtually every other mental and physical attribute were passed through the generations and were heritable along predictable, probabilistic lines. An intelligent father and mother were likely to have intelligent children, he believed, while parents of below average intelligence were unlikely to produce geniuses. 'I conclude that each generation has enormous power over the natural gifts of those that follow, and maintain that it is a duty we owe humanity to investigate the range of that power, and to exercise it in a way that...shall be most advantageous to the future inhabitants of the earth', Galton wrote.[24]

Along with encouraging the increased reproduction of society's most 'fit' individuals through regulated marriages or financial incentives, eugenicists soon realized that their goal of biologically improving human society also required discouraging reproduction of the 'unfit'.[25] Since the mid-nineteenth century, prophets of social doom had been predicting collapse from the supposedly increasing numbers of degenerates in Britain's urban slums, and the country's drawn-out involvement in the Boer Wars led to questions about whether British men were still biologically fit enough to fight. In 1906 and 1907, the government passed legislation allowing for the medical inspection of schoolchildren and provision of basic nutrition in an effort to improve the nation's health and fitness.[26] The eugenicists had a ready-made explanation for this alleged biological decline: the dysgenic (negative, the opposite of eugenic) reproduction of the least fit classes of society in the slums, while the most fit artificially curtailed their reproduction with birth control, moved to the colonies, or simply refrained from marriage.[27] The fit members of society had to therefore be encouraged to reproduce in greater numbers, while the unfit – a category ranging from slum dwellers, the 'morally defective', and chronic paupers, to the disabled and 'feebleminded', depending on who was defining the category – had to have their reproductive rates curtailed. Eugenicist E.J. Lidbetter estimated in the late 1920s that 10 per cent of the British population amounted to a 'Social Problem Group' of intergenerational paupers, the hereditarily disabled, incorrigible alcoholics, inveterate criminals, prostitutes and other irredeemable types, mostly concentrated in the East End of London. Not only was the Social Problem Group trapped in poverty by its biological condition and tainted inheritance, it was also remarkably fertile and internally breeding at a worryingly high rate.[28]

Physical gender segregation and institutionalization was one possible solution for the 'feebleminded' and the disabled, but more extreme measures might be required to deal with the Social Problem Group on a larger scale. By 1910, eugenicists in the United States were pioneering an innovative solution: surgical sterilization. In 1907, the State of Indiana passed the first legislation allowing officials to order sterilization for individuals presumed to be carrying a hereditary malady. Two years later, California passed an expansive law that made the state not only a model for eugenicists around the world but also the most aggressive sterilizer of its own citizens before the ascension of Adolf Hitler.[29] In 1927, the US Supreme Court ruled in the case of *Buck v. Bell* that compulsory sterilization was constitutional and did not amount to cruel and unusual punishment.[30] By 1931, half of the American states had sterilization laws of some form on the books.[31]

Eugenics took a much slower course in Britain. In 1907, a group of eugenics enthusiasts formed the Eugenics Education Society in London to further their ideas through public education and propaganda.[32] Galton was worried that the Society would become a refuge of unscientific fanatics and he initially declined to be involved in its activities, though he eventually recanted.[33] In 1911, Charles Darwin's youngest son, Leonard, became president of the body, increasing its establishment credibility. The Eugenics Society, as it would rename itself in the late 1920s, soon became the institutional home for most eugenicists in the country, though it was never particularly stable and often beset by internal divisions.

Pitt-Rivers had obvious sympathies for eugenics that had begun at an early age. His initial interest in the field was sparked as a student at Eton when he read an abridged version of Rev. Thomas Malthus' *An Essay on the Principle of Population* that he later reported had 'stirred my eugenic conscience'.[34] In perhaps the most famous work of demography ever written, Malthus argued in 1798 that in times of abundant resources, particularly food, human populations inevitably grow until all available resources are consumed – a 'Malthusian Catastrophe' – at which point society is plunged into shortage, famine and collapse. Once this takes place, the population shrinks through death and restriction of reproduction until there is again a surplus of resources and then the cycle begins anew.[35] For neo-Malthusians of the early twentieth century, poverty was therefore seen as the result of overpopulation, and consequently a falling birth rate, which Britain had seen since the 1880s, was a welcome and natural check against catastrophe.[36] The eugenicists, however, were concerned with the social groups in which the birth-rate decline was taking place, fearing that the 'unfit' had not curtailed their reproduction to the same degree as the fitter classes of society – indeed, it was well documented that rich families were usually smaller than poor families – and as a result the country might soon be facing a flood of degeneracy unless countermeasures were taken.[37]

Pitt-Rivers' aristocratic inclinations inclined him to share this latter view, and Galton's focus on breeding the best in society also fit well with his Nietzsche-influenced belief in the need for a robust upper class to combat social decline and Bolshevism. His argument in *Conscience and Fanaticism* about degeneracy causing the outbreak of the First World War had been fundamentally rooted in the idea that increasing numbers of the hereditarily unfit – coupled with the influence of religion and political empowerment of the masses by democratic reforms of the nineteenth century – had been responsible for the conflict. *The World Significance of the Russian Revolution* was likewise predicated on the notion that the world's Jewish community had been collectively responsible for the overthrow of the Tsar and the rise of Bolshevism whether it was aware of it or not, using Freudian psychoanalysis as the explanation. *The Clash of Culture* had focused on the destructive influence of white interference on native marriage customs and argued that intermarriage between races was destructive to both groups and potentially fatal to the 'lesser' race. Urbanization and industrialization of non-whites in Africa, he claimed, had led only to racial antagonism that had given Communist agitators huge numbers of new recruits.[38]

Eugenics obviously appealed these rigid views of race, scepticism towards racial interbreeding and anti-Semitism – combating Bolshevism meant addressing the responsibility of Jews for its spread and avoiding the increase of Jewish influence by interbreeding – but it also appealed to Pitt-Rivers' iconoclastic outlook towards religion and conventional morality. Galton himself had been no friend of Christianity, arguing that religious views of marriage and reproduction were misguided and led to dysgenic outcomes by encouraging all classes of society to procreate and discouraging state interference in home life. In his more extreme writings, Galton even suggested that eugenics should eventually strive to replace the church by creating an ersatz religion that demanded unquestioning obedience to its tenets.[39] It was no accident that eugenics found many of its early converts among social radicals and sexual reformers.[40]

Pitt-Rivers had little in common with the radicals, some of whom were critics of Britain's class structure and the aristocracy itself, but he did believe strongly in reforming conventional views of sex and procreation. He cultivated a friendship with radical sexologist and eugenicist Havelock Ellis, who had scandalized Edwardian Britain with his frank writings about sexuality and 'sexual deviations' by arguing that individuals should be free from religious and social constraints to engage in any sexual behaviour they wished, including homosexuality.[41] Ellis was a sceptic of the Eugenics Society on the grounds that it was ineffectual, telling Pitt-Rivers, 'it is, I fear, too true that the E.E. [Eugenics Education] Society is made up of incompatible elements and so it has no policy'.[42] In 1929, Pitt-Rivers attended Ellis' 70th birthday party with Malinowski and later argued that Ellis' views towards sex were in line with his own and supported the Freudian theory that unnatural sexual repression led to violence:

It is both instructive and significant to remember the obstinate resistance and boycott Ellis encountered in England when he broke the silence that hedged about the domain of marital relations and the sexual life of 'civilized' people, so long the close preserves of orthodox medicine and theology. A prejudice that even now has authoritarian, if not such open, backing, and has its still graver reflections upon the conflicts and mal-adaptations recognized only when official classification distinguishes … the alarming prevalence of the insanities, neuroses and mental deficiencies – the mental casualties of war and social chaos.[43]

Pitt-Rivers joined the Eugenics Education Society as a life member in 1919 during his first period of academic research at Oxford. He was clearly sympathetic to the Society's overall goals but considered its leadership too soft and unproductive for his tastes. He complained that the organization was not sufficiently interested in racial issues, claiming that eugenics required building racial self-consciousness and preserving the purity of the English race in the face of Jewish attempts to attack the very notion of race. He sent a scathing letter to the Eugenics Education Society's leadership soon after joining, in which he accused it of neglecting racial issues:

> Because there are already many racial strains present in the composition of the English population, is that a reason for entirely obliterating what distinct strains still remain by still further dilution? Is racial self-consciousness of no value as a moral and eugenic agency? It would seem that while Jewish publicists everywhere insist (among themselves) upon their racial distinctiveness, they are equally anxious to obliterate by all other adulteration racial types in order eventually, we suppose, to achieve two types only in the world – *homo Judaicus* and *homo vulgaris?*[44]

This was little more than his standard anti-Semitic line – that 'Jewish publicists' had an interest in deprecating the idea of racial purity for other groups while privately insisting that the Jewish race remain pure – but it is telling that by 1920 Pitt-Rivers had already linked these views to his belief in the power of eugenics. While the idea of eugenics had not featured directly in *The World Significance of the Russian Revolution*, it would become an increasingly important aspect of his work after the mid-1920s.

In addition, these concerns over 'racial self-consciousness' and the Jews were echoed by Pitt-Rivers' belief that the society lacked a definite direction and consisted of incompatible elements ranging from left-wing socialists to right-wing anti-Semites like himself. Despite his obvious conflict with the organization's leadership, by 1920 he had been elected to the Eugenics Education Society's council, giving him a key role in its governance. This was not as unusual a decision for the society as it might seem. In the words of the council itself, Pitt-Rivers had established himself as a 'sincere and sound eugenist' whose eccentricities might be outweighed by what he could potentially be able to bring the organization in both money and new recruits.[45]

There was also one topic where Pitt-Rivers found common ground with some of the more socially radical elements of the Eugenics Society: birth control. This was a supremely controversial topic in British society in the 1920s generally and within the eugenics movement itself. Since the

1880s, it was widely known that well-to-do couples had enjoyed access to increasingly reliable contraceptive knowledge and devices, reducing family sizes dramatically (as an illustration, Pitt-Rivers' own parents had far fewer children than his grandparents). There was widespread concern over what this might mean, both morally and medically, for the country and the empire. At the 1908 Lambeth Conference, the Church of England had denounced contraception as both dangerous and immoral, and British medical journals debated the safety of various birth-control methods and devices throughout the early years of the century.[46] Opponents argued that the widespread use of birth control would lead to the uncoupling of sex and marriage from reproduction and result in widespread sexual wantonness. Some eugenicists feared that if access to contraception was increased, the 'wrong' people (the middle and upper classes) would be the most likely to use contraception while the poor and the Social Problem Group would continue to reproduce at the same rate.[47] Others, including Ellis, disagreed, and argued that the use of contraception by the middle class led to the emergence of smaller, healthier families that could afford the necessities of life and hope to maintain their social position.[48] This debate over contraception and its demographic consequences raged in the Eugenics Society throughout the first half of the twentieth century, and in 1936, Leonard Darwin diplomatically wrote that, in his view, increasing access to birth control might carry both positive and negative consequences depending on the couples that decided to use it.[49]

By the early 1920s the British birth control movement was led most openly by Marie Stopes, a former palaeobotanist who began her career studying ancient fossils and impressively earned PhDs from both the University of London and the University of Munich. In 1911 she had unwisely married Canadian botanist Reginald Ruggles Gates, who would later become a friend of Pitt-Rivers and have a lengthy and controversial career as a racial researcher.[50] It is not entirely clear what transpired in the confines of their marital bedroom, but Stopes later claimed that her marriage to Gates had never actually been consummated.[51] She eventually filed for an annulment on these grounds and a medical examination allegedly revealed that she was still a virgin at the time of the legal proceedings. Gates contradicted the claim, testified that they had regularly had sex using a variety of contraceptive methods and presented a medical document stating that he was not impotent. At the same time, he did not contest the annulment.[52]

Regardless of whether she had ever actually had sex, Stopes now embarked on a campaign to save couples from the sexual ignorance that she claimed had doomed her marriage, intently reading the works of Ellis and other sexologists in the reading room of the British Museum (the same room, incidentally, in which Karl Marx had researched economics decades before). In 1918 she published *Married Love*, a shockingly explicit sex manual for the time that emphasized female pleasure and advocated the use of contraception to build strong families.[53] Conservatives were scandalized, while reformers were thrilled by her frank discussion of female sexual pleasure. As Clementine Churchill wrote to her husband Winston, 'I can't think why this pamphlet was not written years ago ... I hope it will not be jeered at by middle aged, plain cynical men over their wine. It is meant for beautiful god-like young men when ardently pursuing tender and lovely nymphs.'[54]

With this controversial emergence in the public consciousness, Stopes soon became one of the leading contraception advocates in Britain. Her interest in sexual education included a eugenic bent, and her advice to working-class women focused more directly on birth control devices and methods and less on female pleasure than her advice to the middle and upper classes.[55] She later disowned her son for marrying a woman who wore glasses and thus presumably carried poor hereditary material.[56] The cover of *Birth Control News*, a journal published by the Stopes-led Society for Constructive Birth Control and Racial Progress, boasted on its cover that 'deliberate motherhood' was 'a sure light in our racial darkness' and its pages carried sensationalist stories about the number of mentally deficient children being born in Britain.[57]

Stopes was too radical for many of the 'respectable' members of the Eugenics Society and Leonard Darwin treated her with particular contempt. She was allowed to join the organization in 1921 but was not permitted to use its official channels to promote the Society for Constructive Birth Control out of fears that she would alienate anti-birth control eugenicists.[58] Pitt-Rivers soon became one of her leading and most vocal supporters within the society's leadership. His own view of birth control was that it should be extended as widely as possible to encourage the rearing of small but eugenically fit families. 'It must be obvious to all who think that the birth-control movement at present has the effect of sterilising the wrong elements in the population: there is not even the remotest chance of it ever being abolished, the only possibility of mitigating its dysgenic influence is therefore to extend it so that its influence may be eugenic instead of dysgenic', he wrote.[59] Pitt-Rivers subsequently addressed a Society for Constructive Birth Control meeting in 1932.[60] Stopes, for her part, considered Pitt-Rivers to be an important ally and asked him to review a book denouncing Catholic birth control methods.[61] He complained to her in 1934 that the society's members would too often 'for the most part go to the ground like the rabbits they are' when birth-control issues came up in committee meetings rather than risk a confrontation with other members.[62]

This increasing frustration over the Eugenics Society's general conservatism was a recurring theme for both, and by the early 1930s, Pitt-Rivers believed that the society's wishy-washy approach to eugenics had allowed the term to be associated with any range of positions including religious perspectives that he found antithetical to a 'scientific' mindset. As he had written in a scathing letter published in the *Eugenics Review* in 1920:

> I feel perfectly sure that every individual eugenist, *qua* eugenist, has nothing to fear from opponents. As a 'eugenist' – if a small yearly subscription to the Eugenics Education Society and an incorrigible fondness for the study of the human animal gives me a right to use that title – I should like to feel equally confident that the Eugenics Society is invulnerable and its policy unquestionable …
>
> The Society has, I am willing to believe, a definite policy, based on definite first principles, for the purpose of achieving a definite eugenic ideal. How then are we to reply to critics (including even distinguished members of the Society itself) who assert that a Society composed of people holding mutually incompatible views, founded upon opposed principles and values, could not and does not put forward a policy sufficiently consistent to achieve anything worth while?[63]

Interventions of this type would gradually make Pitt-Rivers powerful enemies within the Eugenics Education Society, but throughout the 1920s he enjoyed increasing international recognition. In 1929, he became the society's official representative to the International Federation of Eugenic Organizations (IFEO) on the personal nomination of Leonard Darwin.[64] The IFEO had been founded to encourage international connections between eugenicists in various countries and their respective organizations by holding regular conferences and other events.[65] Pitt-Rivers was probably nominated as the society's IFEO representative because of the copious amount of time he could dedicate to conference travel and other responsibilities, along with the financial resources he could use to support its efforts.[66]

The first IFEO event Pitt-Rivers attended in 1929 was held in Rome and attended by many of the leading names in the international eugenics movement, including German racial hygienist Eugen Fischer, the director of the Kaiser Wilhelm Institute of Anthropology, Human Heredity and Eugenics, and American eugenicist and anti-immigration campaigner Harry H. Laughlin. In addition to the usual scientific papers, the event included the issuance of a special commendation for Mussolini, who Fischer claimed had done 'best of all rulers in regard to population problems – and further his is the only Government which can undertake practical legislation on such points as the differential Birth-rate. It would be a matter of world-wide significance if Italy made a move in this matter'.[67] Issuing the commendation had been carefully planned in advance in an effort to further the agenda of the eugenicists by courting the *Duce* directly. 'There will never again be such an opportunity to explain our ideas clearly to a statesman again', Fischer told Charles Davenport, one of the founders of the American eugenics movement. 'And he is the only politician who can carry out eugenic measures and perhaps will.'[68]

Enamoured with the IFEO and its activities, Pitt-Rivers now made the decadent gesture of offering to personally hold the next international meeting at his Hinton St Mary estate. This was a major departure from the largely urban venues that had been chosen in the past and it was a huge undertaking: dozens of esteemed scientists from around the world would have to be hosted, entertained and transported from their ports of arrival to the rural conference venue. A number would be hosted in Pitt-Rivers' renovated manor house itself. When it kicked off in September 1930, the Ninth Conference of the IFEO guest list included many of the most prominent eugenicists from all over the globe. Leonard Darwin, Arthur Keith, and Reginald Ruggles Gates, all personal acquaintances of Pitt-Rivers, constituted the British delegation. Representing Germany was Alfred Ploetz, the intellectual progenitor of the German racial hygiene movement, along with Ernst Rüdin, a Munich-based psychiatrist and racial hygienist (Eugen Fischer did not attend, citing illness).[69] Laughlin came from the United States and used part of his trip to investigate British immigration policy.[70]

A newspaper reporter covering the event was amazed at the spectacle of hosting a meeting of such renowned scientists in a Dorset village populated by people the correspondent compared to the characters in a Thomas Hardy novel. Asked about his views of the event, one resident reportedly replied simply 'I don't hold with politics.'[71] The reporter himself was more sanguine, concluding his story with the statement, 'If eugenists have their way – and ultimately I have no doubt they will obtain the ear of statesmen – then a new phase in the evolution of mankind will be initiated. If we find that the way we are living is leading us straight to physical and mental bankruptcy, then we can no longer afford to be mere pawns on the chessboard of evolution. We must somehow take a hand in the game.'[72]

After days of meetings and deliberations, the eugenicists considered the Dorset event a great success. Excursions had taken them to visit the Pitt-Rivers Museum in Farnham, the Giant of Cerne and Stonehenge. Public talks were arranged for the locals, and Gates was among the speakers calling for a careful tracking of racial crossing around the globe for possible dysgenic effects, while a joint paper by Laughlin and Fischer called for more research into racial questions.[73] Delivering an encomium for Leonard Darwin (calling him the 'Nestor of English Eugenists'), Arthur Keith predicted a strong future for the international eugenics movement and proclaimed that 'the public of all countries is more willing to listen to Eugenic proposals than at any previous period. No doubt this change of mood has been brought about by the activities of societies represented in this international organization'.[74] There was little doubt that the event had been a major accomplishment, and many of the connections Pitt-Rivers cultivated through the IFEO would become significant in later years.

Successfully hosting the 1930 event was clearly a feather in Pitt-Rivers' cap on the international stage but it did little to increase his standing within the Eugenics Society itself. By now the society was in the midst of a major split between a younger reformist faction and the older members of the organization, represented in part by Leonard Darwin and his allies. The reformers were led by Carlos Paton (C.P.) Blacker, one of the most unique and in many ways beguiling figures of twentieth-century British eugenics. He and Pitt-Rivers shared a somewhat comparable background: Blacker had been educated at Eton and was wounded in the First World War after his older brother had been killed.[75] Memories of the war seem to have haunted him for the rest of his life and the conflict led him to abandon his Catholic faith.[76] Following the Armistice, Blacker attended Balliol College, Oxford, where he studied under Julian Huxley.[77] By the 1920s, Huxley was not only a leading zoologist (as noted, he had criticized *The Clash of Culture* on the grounds that it was unscientific) but also a well-known scientific commentator and popularizer. He would soon become Blacker's most significant ally in the difficult political battles within the Eugenics Society. By the mid 1930s, Huxley was convinced that the underlying ideology of eugenics was similar to Nazism and as a result he began to revise his views of both the field and racial issues generally.[78] He would later emerge as a prominent anti-racist voice in the British scientific establishment.

After completing his studies at Oxford, Blacker studied medicine, trained as a psychiatrist and began to practice at Maudsley Hospital in London.[79] Though he disliked Stopes' emphasis on female sexual pleasure and feared the growth of immorality, he allied himself, and later the Eugenics Society, to the birth control movement.[80] In 1927, he publicly argued that physicians alone should be allowed to dispense birth control information and all non-medical information sources should be legally banned to ensure reliability. To make this possible, however, doctors would have to move beyond their personal concerns about contraception and accept its role in preventative medicine, which many had previously been unwilling to do.[81]

Almost always the consummate diplomat, Blacker's approach to eugenics and its political agenda emphasized public moderation. He was sceptical of traditional Galtonian notions of racial and social worth. Instead, he argued, eugenics should focus on political reforms to rationally manage population size and quality, along with the elimination of hereditary disease. The policy he was most eager to embrace was the legalization of voluntary eugenic sterilization, the legalization of which might be used to accomplish eugenic aims through a veneer of volunteerism rather than compulsion.[82] Under this model, surgical sterilization would be legalized for individuals who requested it (most experts agreed that it was presently probably illegal for a doctor to perform a sterilization procedure even if the patient had requested it, though there was some disagreement on this point). The greater difficulty came when the patient requesting the procedure was not deemed mentally sound enough to meaningfully consent, in which case the decision would be left to family members or other caregivers, including potentially the state itself. Blacker believed that if British law could be amended to allow the sterilization of individuals on a voluntary basis, many 'deficients', along with others, would choose to undergo the procedure to avoid the demands of parenthood and thus give eugenics a much-needed political victory. Countries with compulsory sterilization laws were, of course, important models upon which this campaign could in part be based, though the British campaign would always ostensibly emphasize the idea of volunteerism and personal liberty.[83]

In late 1930, Blacker became general secretary of the Eugenics Society, a new position that gave him immense power over the organization and its sizable financial assets.[84] He and Pitt-Rivers initially got on well and shared much in common. They had both attended the same school and university, and both had served in the First World War before being wounded. They were also virtually the same age – Blacker was five years younger – and they both believed in legalizing birth control across the social spectrum. However, the similarities only extended so far. Blacker believed that campaigning for only voluntary sterilization, without a compulsory aspect, and building alliances across the political spectrum was the key to success, with his fellow medical practitioners taking a leading role. As Darwin put it, Blacker's strategy was to 'influence different bodies by direct methods of attack, especially the medical profession, politicians and trade organizations' to build support for eugenic reforms, rather than taking the lead in pushing for them in the public sphere when it might be politically dangerous for the society to do so.[85]

In Pitt-Rivers' mind, this approach amounted to cowardice and defeatism. Galton's original vision for eugenics, he believed, was for it to become enshrined in the national conscience and take a guiding role in both state policy and personal decisions. His research in the South Pacific had seemingly reinforced the necessity for an international eugenic effort to stave off imminent racial decline. This was an urgent matter, and the Eugenics Society was already too sensitive to opposition opinion, in his view. He confided his frustrations in a letter to Reginald Ruggles Gates:

> Has the Eugenics Society any collective scientific conscience? Its founder Galton being a psychologist as well as a scientific methodologist was keenly attuned to the need for appealing to intelligence rather than prejudice and ignorance, consequently he would never allow considerations of popular propaganda to outweigh a conscious

adherence to scientific veritas so far as they were then known. We seem
to be much more concerned not to shock members of the Y.M.C.A., or
the Catholic Church or Methodist Church, or Fleet Street journalists
or the anti-vivisectionists, or the Labour Party intelligentsia or any other
group who know the who truth straight from God or Karl Marx or Mrs.
[Mary Baker] Eddy or Mr. Stalin, than we are to shock any scientific
group of workers in genetics, anthropology or any other discipline.[86]

Thus outraged by the spinelessness he sensed, in 1931, Pitt-Rivers
published a scathing attack on its approach to eugenics and on the field's
critics alike. The short book was entitled *Weeds in the Garden of Marriage:
The Ethics of Race and Our Captious Critics* and was ironically – and
harshly – dedicated to 'the mother of my sons' who he presumably held
responsible for producing 'weeds' in his own marriage.[87]

Weeds in the Garden of Marriage was little more than a polemic. Pitt-
Rivers had initially pitched the manuscript to T.S. Eliot, then a director at
Faber & Faber, who rejected it on the grounds that the subject was too
extensive, and the science too complicated, to be adequately addressed in
such a short form. 'That is my opinion; not in spite of, but as much because
of, my own interest in the essay and my profit in reading it', Eliot told him.[88]
The book was eventually published by N. Douglas and opened by defending
eugenics as an indisputably scientific field of study:

> The facts discovered by researches in biology and genetics relate,
> of course, to all living organisms, to plants and animals equally with
> mankind, and so far as any man tries to apply his knowledge and
> experience to improve the stock in his garden or farm-yard, to encourage
> the growth of stronger or healthier plants, or to eradicate weakness or
> disease amongst his cattle or horses, by encouraging the healthiest and
> finest strains, so far is he acting eugenically.
>
> To condemn or criticize such endeavour would be regarded as idiotic
> and mischievous, yet when these ideals, based on greater knowledge and
> a desire to increase happiness, are applied to the human race, a host
> of denunciatory critics, strange as it may seem, have been known to
> spring up.
>
> Eugenics, whatever the nonsense that may be talked about it against it
> or for it, is one of the most important and humanly valuable applications
> of the key science of Anthropology, the science of Man. No movement
> that attempts to do or to teach something positive can fail to meet with
> opposition.[89]

Pitt-Rivers then classed the critics of eugenics into three groups: those who
feared that it would go too far and encroach on personal liberty; those who
feared that it would accomplish too little to make a significant difference;

and those who 'regard it as a challenge to their own comfortable optimism, a profane doubt cast on their trust in God's safe ordering of the best of all possible worlds' – in other words, those who saw it as an admission that the world was not already perfecting itself either through religion or social change.[90]

He reserved most of his ire for this final group, in which he included figures as varied as socialist author George Bernard Shaw ('Mr. Shaw's notion of Eugenics … is that some busybodies wish to set up a government department to decide how many types of human being are desirable … the answer, of course, is that no intelligent Eugenist has advocated anything so stupid as a Socialistic government department') and Catholic writer G.K. Chesterton ('he really knows no more than Mr. Shaw does about Eugenics – which is very little indeed').[91] Chesterton's religious views were subjected to particular ridicule. Traditional religious and civil prohibitions on incest and inbreeding, Pitt-Rivers claimed, were unscientific, and 'it is far better that defective stock should intermarry and make apparent recessive defects, the more closely related the better, than mate with normal stock. Also the best stock should choose their like, even within their own kinship'.[92] Catholic opposition to birth control and sterilization on the grounds that both were 'unnatural' were disingenuous, he went on, because the Church's insistence on clerical celibacy was a greater effort to 'subjugate nature', as was the past practice of castrating choristers (castratos) to prevent the changing of their voices in puberty.[93]

Having thus excoriated both left-wing and Catholic opponents of eugenics, Pitt-Rivers now launched an ill-advised extended attack on 'one of my respected friends, who happens to be an anthropologist of distinction': Berkeley professor Robert H. Lowie. Two years earlier, Lowie had written a book entitled *Are We Civilized?* that had branded the American eugenics movement as 'a cloak for Know-Nothingism' and racial bigotry.[94] Lowie went on to argue that anti-immigration eugenicists 'talk twaddle about the children of mixed marriages between Europeans and Jews belonging to the lower type – the Jews' and was scathing in his assessment of Galton's work itself:

> Heredity hypnotized Galton as the portent of the stars had hypnotized [Johannes] Kepler. It lulled his critical faculty. The Age of Pericles [in ancient Athens] was a period of marvellous achievement; hence it had to be explained in racial terms. By partly unconscious selection Athens 'had built up a magnificent breed of human animals'. … As scientific proof this is from beginning to end tommyrot. No one *knows* anything about any partly unconscious selection in Athens … All this seems obvious. Why, then, did Galton not see it? He did – when the holy madness was not upon him … Science has made advances; the scientist is still a primitive man in his psychology.[95]

As Pitt-Rivers uncharitably described it, Lowie's argument was 'that all eugenists are ignorant fools; a man who is ignorantly foolish is probably a eugenist or at any rate like a eugenist; therefore Eugenics is ignorant folly, which is a very bad syllogism'.[96]

By specifically evoking the example of Jewish immigration as an example of this folly, Lowie could hardly have constructed an argument more likely to raise Pitt-Rivers' ire. Conceding that anti-immigration sentiment was at times 'couched in terms that are often as exaggerated as they are irrational and emotional', Pitt-Rivers retorted that the Jewish community itself supported his view of eugenics:

> One of the most formidable irrelevancies that is openly and surreptitiously dragged in to obscure the eugenic question is the Jewish question.... The true anthropological point of view is neither 'Semitic' nor 'anti-Semitic,' but must take cognizance of the phenomenon of 'anti-Semitism,' as it must of the equally patent fact that the history of Christendom is the history of the Judaizing of European history.... The Jewish question is not only a problem of culture, it is also a racial problem even more obscure than the culture problem, which is at least one reason why racial problems are habitually so perversely ignored or confused even by anthropologists.... Themselves the most race-conscious people in the world, with a religion and a cultural tradition steeped in racial ethics, Jewish anthropologists have been particularly apt to advertise their origin in their writings by waging an irascible war against all signs of race consciousness and racial ethics in the rest of the world.[97]

In other words, Pitt-Rivers was claiming that the experience of diaspora had created a race consciousness among the Jewish community that was a source of its cultural strength and survival. 'Thanks to a rigid code of exclusive intermarriage amongst themselves, as much as to the prejudice of their neighbours, they have lost from time to time only a small portion of their blood by marriage with non-Jews', he wrote.[98] Lapsing once again into his usual choice of conspiracy theory, he argued that this biological strength was being used as a weapon against the wider world:

> Failing to persuade the rest of the world to accord to them openly the privileges due to a 'Chosen Race', they have at least succeeded in dissuading most of the world from thinking that any other race could be superior, if indeed any other 'race' could exist at all. Failing superiority they have fought for toleration on the grounds of sameness and equality. Instigating and encouraging the confusion that exists between culture and race, and between race and nationality, they have accepted the formula that describes a Jew as an 'English or German or Russian gentleman of the Jewish persuasion.'[99]

Pitt-Rivers thus saw eugenics as the ability to counteract these Jewish influences. Control of human heredity, he claimed, offered the solution to a vast array of problems and issues including 'economics and taxation, public health and housing, crime and the penal code, alcoholism, marriage laws, rural and urban migration, poor laws, immigration and emigration, the Established Church, education, agriculture, industry, employment, labour, colonial administration, the constitution and the social system'.[100] The book concluded with a sixteen-point 'practical program' for 'promot[ing] the welfare, health and happiness of nations and realize an ideal of racial improvement' that included compulsory eugenic education in schools; sterilization laws 'to prevent racial deterioration'; the mandatory exchange of hereditary health certificates before marriage; expansion of divorce laws to allow marriages to be dissolved on the grounds of 'hereditary or chronic insanity, feeble-mindedness, desertion and sterility'; legalized contraception for all women; legalized abortion 'in the interests of either the mother or the health and soundness of the child'; and tax reforms to financially reward child-rearing.[101]

This radicalism was hardly the image of eugenics Blacker was eager to cultivate in the Eugenics Society of the early 1930s. *Weeds in the Garden of Marriage* quite obviously had little purpose beyond provoking trouble within the society and with its opponents. Ironically, Pitt-Rivers had asked Blacker to edit the proofs of the book and presented him with a Japanese vellum copy when it was finished. No doubt on Blacker's urging, however, the Eugenics Society subsequently declined to officially endorse the work and refused to include it on its lists of recommended reading.[102] Opinions among some members were more positive, however, reflecting the emerging split in its membership. Statistician R.A. Fisher, a major voice in Darwin's traditionalist faction, saw little objectionable in its content: 'I may say that the criticisms that it is liable to shock the timid, or antagonise the suspicious, or that it is democratically unpalatable, have not in fact come my way. I have heard nothing worse than "confused" and "sensational"', he told the author.[103] Sending a copy to Lowie, Pitt-Rivers attempted to gloss over his scurrilous attack, telling the Berkeley anthropologist that he hoped he would 'take [it] in the generous and fraternal spirit that you showed me in California – my chiding was I must confess primarily provoked by your polemic against those ineffably stupid "Eugenists"!'[104] This effort was met with a stony response, and their relationship seems to have been more or less ended by the publication. It would be only the first friendship that Pitt-Rivers would sacrifice in the name of his polemic insistence on always being right.

For their part, academic reviewers were generally positive towards the book. Berkeley zoologist Samuel J. Holmes praised the work in the *Journal of Heredity* as a strong defence of scientific ideals and for revealing that 'anthropology, for some reason, has come to be largely a Jewish science, and most of its voteries [*sic*] have a very obvious bias toward egalitarianism.

Capt. Pitt-Rivers, who is himself a professional anthropologist, is one of the few who have commented upon this rather interesting anthropological situation'.[105] A review in *Nature* was even more openly adulatory. Calling Pitt-Rivers 'intensely in earnest' and completely ignoring his sections on the Jews, the reviewer wrote that the book was an argument for 'seeing with him [Pitt-Rivers] that a thoughtful application of the principles of eugenics to our economic and race problems is our only salvation... he sees the race, indolent, ignorant, and vain, using its power of reproduction to destroy itself.... Make the world safe for intelligence, is Mr. Pitt-Rivers' cry, and all good things will be added unto you'.[106]

Within the Eugenics Society, however, Pitt-Rivers' irascible personality and increasingly extreme views were causing more problems even before *Weeds in the Garden of Marriage* appeared in print. In late 1931 he 'damned nearly came to blows' with a fellow Council member after being treated rudely during a meeting.[107] Blacker personally vetoed a proposal to host a debate on the topic of 'religion and eugenics' between Pitt-Rivers and Chesterton on the grounds that 'G.K. Chesterton would mop the floor with him – make a fool of him and cause him to lose his temper. When he does this, he bellows. Ghastly.'[108] The debate eventually took place, but not with Chesterton as the adversary. In it, Pitt-Rivers pretentiously proclaimed that 'Christianity might be increasingly converted by eugenics, though eugenic ideals could not be changed by Christianity'.[109] Few members of the society were impressed.

Outraged by the Eugenics Society's failure to endorse his book and its rejection of his debate proposal, Pitt-Rivers now launched a direct attack on Blacker's leadership. In March 1932, he sent a provocative multi-page memorandum to the council of the Eugenics Society accusing the body of failing to adopt the necessary policy stances to achieve political success and adopting a 'safety first' policy that he had deliberately violated by publishing *Weeds in the Garden of Marriage*. The timid stances of the organization had ruined its past chances of success: 'The Society has lost far more than it has gained in influential support through a too timorous desire to avoid controversy and conciliate all views than by any opposition we have met or provoked', he claimed.[110] Echoing his sentiments from more than a decade earlier, he argued that the society should publicly stake out deliberately controversial stances on contraception, sterilization and possibly even elective abortion in an effort to lead public opinion in its direction rather than pursue conciliatory policies towards opponents.

Circulating his memorandum to key members of the society, Pitt-Rivers received a number of supportive notes, many of which came from older and increasingly isolated members of the society who believed that Blacker's leadership was squandering Galton's legacy. Havelock Ellis, who had himself clashed with Blacker, praised Pitt-Rivers for trying to 'infuse vitality into the Eugenics Society' and penned a scathing accompanying note:

There can be no doubt whatever that the Society has lost far more support than it has gained by the timorous desire to avoid opposition and controversy and meekly to placate the forces of inertia. I am constantly hearing the Society referred to as having become effete and reactionary (which no one in old days ever said of Galton), so that I am now rather nervous about having my name associated with the Society. There is a widespread feeling that the cause of Eugenics calls for the formation of a new Society to take it up and carry it forward in a more intelligent a fearless spirit. Personally I would prefer, if that is possible, to see a change of heart in the Society itself. No doubt it would need some weeding out. But if a Eugenics Society is unable to deal with its own unfit, how can it claim to deal with the unfit of the world at large?[111]

From his position as perhaps the most prominent religious advocate of eugenics in the country, Dean William Ralph Inge of St Paul's Cathedral praised the sentiment of seeking a more aggressive stance against Catholicism, though he disagreed with the notion that younger members of the society were less confrontational than Galton would have liked. 'The Society, not too numerous, must be kept together. If it declared itself in favour of (say) companionate marriage or the legalisation of abortion...the Society would lose useful members. You do not want to have the whole religious world against you', Inge wrote.[112]

The leaders of the Eugenics Society were far less amused. Blacker wrote to Pitt-Rivers asking for clarification ('Are you able yet to focus the general ideas you set forth in your memorandum upon any practical proposal? That is to say, are you contemplating revising our statement of policy in regard to birth control, and do you wish us to express approval of the legalising of abortion?') and society President Bernard Mallet conceded that while 'we have become a good deal more progressive in the year or two since Blacker has been with us', he argued that the society's political strategy was fundamentally sound.[113] The society's rejection of *Weeds in the Garden of Marriage* was based on similar logic, Mallet wrote:

As regards your book [*Weeds*], it seemed to me very good on eugenics, but there certainly was a feeling in the Committee that it might still further alienate Labour, which as you know is one of our chief difficulties. It was therefore decided not to recommend the book officially, though of course we bought some copies for the library. I agree that this was rather a timid proceeding and I ought perhaps to have pressed more strongly than I did to have it regularly adopted.[114]

Refusing to give up, in a follow-up memorandum the following month Pitt-Rivers provocatively reiterated his argument that the society's timid stances were leading it to failure. 'It is not a quarter of a century since

Francis Galton declared that "the fit moment to declare a 'Jehad' or holy war against customs and prejudices that impair the physical and moral qualities of our race will be when the desired fullness of information shall have been acquired"', he wrote. 'I should be glad to move that this Council be asked to determine whether we are, after this long interval, any nearer that moment.'[115]

Despite the intensity of his appeal, the council declined to take meaningful action on these documents and proposals. Frustrated by yet another painful rejection, Pitt-Rivers now looked beyond the society for support. He suspected that the problems he had diagnosed within the organization were actually endemic to eugenics as a discipline. The solution to the intrinsic problems with eugenics, he believed, was to introduce more vigorous 'anthropological' notions of race and culture into its discussions, particularly in its considerations of 'primitive races' and the Jews. The result would be a new field he called 'Ethnogenics' that he believed could combine the insights of anthropology and eugenics to study the phenomenon of racial decline and extinction.

Initially proposing the discipline's creation in a paper delivered in 1931 and published the following year, Pitt-Rivers pretentiously described it as a fusion of racial science and the eugenicists' emphasis on propaganda and political action:

> Human Biology thus has a concern in all the influences, cultural, physical and environmental, that affect the extinction or the survival of ethnic variations or types.... Ethnogenics is therefore the study of those forces, amenable to social control, which may influence the fertility and survival rate of variations of type in a population: it is a necessary aspect of that 'functional anthropology,' which has two distinguished exponents in Dr. Malinowski and Professor Radcliffe-Brown. We are often reminded that the backward aboriginal races of the world are fast disappearing, as an urge to speed our investigations of their cultures before our material is lost forever, but ethnic types may disappear amongst thriving and dense populations, even here in our midst. Are we less urgently concerned in preserving those types and those races which enrich our civilization than in writing with academic precision, the obituary notices of those most remote from us; or are eschatology and the measurement of skulls the only really important branches of Anthropology?[116]

Two years later, Pitt-Rivers reprinted this paper in a volume honouring C.G. Seligman. The second version was nearly identical to the first but with a tellingly modified paragraph towards the end that connected Ethnogenics directly to his political views:

> Though perhaps less conspicuous than in the new world of America, there are evidences of social disintegration in England, in perhaps a different

form, symptoms of the same disease that destroys societies and brings about the collapse of civilizations and cultures. A growing consciousness that the existing political and economic structure of society is ceasing to function and is bankrupt, while it breeds disillusionment on the one hand, has fostered on the other a new determination and hope, expressed in the aspirations of various nationalistic Fascist movement in Europe, that may ripen into a new Renaissance.[117]

With his creation of Ethnogenics, Pitt-Rivers believed he had placed eugenics on a scientific and political footing that would make it not only more intellectually rigorous but also effective in creating the racial consciousness that he believed was lacking in the mainstream of the movement. In addition, however, he also believed that it could serve as the scientific underpinnings of a future fascist state and a 'new Renaissance'. The fact that he was the only scientist to ever embrace Ethnogenics or apparently take it seriously as a field seemed to only increase his dedication to it.

Now, effectively marginalized from the Eugenics Society, Pitt-Rivers focused his intellectual energies on another body that seemed more receptive to his ideas. In 1928, he had been involved in the formation of the International Union for the Scientific Investigation of Population Problems (IUSIPP), an international organization dedicated to population and demography issues more generally. As the organization's president, Sir Charles Close, noted in 1931, 'the Union desires the assistance of all those men and women of science, all over the world, who are disposed, by joining its counsels, to help in the work of the scientific examination of the many difficult and pressing problems of population'.[118] The remit of the IUSIPP was therefore much broader than that of the Eugenics Society or the IFEO and focused on a wide variety of population-related issues around the world.

Organizationally, the IUSIPP was composed of national committees elected from its member countries. These committees paid annual dues to maintain their affiliation and the organization's international conferences were held in a different venue each time they took place. The body's executive committee consisted of a president, a number of vice presidents, and an honorary general secretary who also served as treasurer. The final authority on decisions lay with the General Assembly of delegates from the national committees. In addition to its conferences and other meetings, the organization published a journal called *Population*. American Raymond Pearl, a friend of Pitt-Rivers, was elected as the union's first president, while Pitt-Rivers himself initially served on a commission examining 'vital statistics of primitive races' that was chaired by Malinowski.[119] At the IUSIPP's 1931 general assembly, Pitt-Rivers was elected as honorary general secretary and treasurer. He later edited and published the proceedings of the event, which received widespread praise from the leaders of the organization.[120]

British membership in the IUSIPP was maintained through a body known as the British Population Society (BPS), which was funded in large part by money from the London School of Economics and its director, Sir William Beveridge. In 1932, Pitt-Rivers was elected Chairman of the BPS on Close's direct recommendation. In justifying his nomination, Close told Beveridge that Pitt-Rivers 'will be able to give a good deal of time to the work...he has plenty of energy'.[121] It was certainly true that Pitt-Rivers had time to spare and plenty of energy, but Beveridge and Close would soon find that he would not necessarily apply either to the work of the IUSIPP in the ways they hoped. Pitt-Rivers' status within the international organization and the BPS gave him a platform from which he would soon try to push both into line with his political views, with difficult consequences.

By the early 1930s, Pitt-Rivers had thus been rejected from the Eugenics Society for his extremist views and had moved on to what he believed would be a friendlier organization. As the decade went on, however, it quickly became clear that his interests lay less in science and more in the promotion of his anti-Semitic views and those of his allies. Close and IUSIPP would soon regret placing him in a position of responsibility while Blacker and the Eugenics Society would be pleased that they had marginalized him before more serious harm could be done. Bronislaw Malinowski and Oscar Levy, on the other hand, would soon become increasingly concerned that their old friend was heading in a direction that could only lead to his destruction.

CHAPTER FIVE

Becky and the Nazis

By the early 1930s, George Pitt-Rivers' academic credibility had already been significantly frayed. His public embrace of eugenics with particularly racialist undertones had earned him the ire of many within the Eugenics Society, most notably C.P. Blacker. The fact that Blacker was clearly seeking to force him from the society's leadership by the mid-1930s and restricting his public appearances demonstrates the level of animosity that he had already generated in these circles. On the other hand, he remained in a position of authority within the IUSIPP that allowed him a platform to espouse his views at international conferences and other gatherings. While the Eugenics Society closed its doors to him, Pitt-Rivers embraced his role in the IUSIPP and evoked its authority in his increasingly questionable scientific pursuits.

By mid-1933, the international situation had begun to change dramatically as well. In January, Adolf Hitler had become chancellor of Germany, giving his National Socialist Party its first taste of national power though it lacked a parliamentary majority in the Reichstag. On 27 February, a Dutch communist named Marinus van der Lubbe entered the Reichstag building after nightfall and set the building alight. The flames spread to the main chamber before firefighters could extinguish them, destroying much of the building. Evoking the spectre of imminent communist revolution, Hitler urged President Paul von Hindenburg to sign an emergency decree suspending most civil liberties. The Communist Party was subsequently banned and its leaders were arrested. In March, the Enabling Act allowed Hitler to effectively circumvent the constitution, putting Germany on the road to becoming a one-party state.[1]

The rise of Hitler, and his aggressive response to the Reichstag Fire, impressed Pitt-Rivers. The fire itself seemed to confirm his long-standing suspicions about an international communist conspiracy, and Hitler's anti-Semitism corresponded with many of the views he had espoused in *The*

World Significance of the Russian Revolution more than a decade before. 'If the Fascist Revolution had not occurred in 1922 in Italy, it might, we are entitled to speculate, have first occurred in England; for the ideas and the forces which have expression to what is now called Fascism were in the womb of time and gradually shaping in men's minds many years, even many generations, ago, in England, Australia, America, Italy, Germany, and France', he told a German crowd in late 1934.[2] He also possessed a deep personal admiration for Hitler, reflecting after reading his autobiographical *Mein Kampf* that 'the strongest impression left upon me by the man is his immense courage and will to fight, based on a very deep and long vision – the two essential qualities for leadership'.[3]

This affinity for Hitler and Mussolini would soon come at the price of Pitt-Rivers' academic legitimacy. His relationship with Malinowski was already past its prime. While the anthropologist and his children visited Pitt-Rivers' estate in 1933, they corresponded little afterwards, and Malinowski soon blamed himself for the deterioration in their friendship:

> I was thinking the other day, and very often in the meantime, how rapidly time is flying and how you and I are just becoming two old fogies and imbeciles (especially I, of course), and getting constantly out of touch with one another. I am getting so rushed and rheumaticky and soft-headed and bad-tempered, that it is impossible for me to do anything but swear and write business letters. So that the present, which is neither of these two activities, may prove to you that you are still very often in my mind, thought, emotions and conditioned reflexes.... In the spring we must meet again. Give my love to your fair Lady and take some of it for yourself.[4]

Malinowski's next note announced the death of his wife, Elsie, and expressed 'my sincere and unchanged friendship and the hope that when you come next to town you'll let me know and we meet quietly for a long talk. I have grown quite old in that I prefer to think of the past and am really tired of life'.[5] Malinowski would die abruptly in the United States in 1942 and there is no evidence that he and Pitt-Rivers had been in contact in the years immediately before.[6] No doubt one aspect of this breakdown in their relationship stemmed from their increasing political differences. By the mid-1930s, Malinowski was publicly denouncing Nazism as a form of 'modern magic' that would lead only to senseless bloodshed and serve no positive social function. Following his death, his surviving second wife published a book entitled *Freedom and Civilization* based on his lectures that amounted to a direct denunciation of Nazi Germany. The fact that Pitt-Rivers only referred in his later writings to Malinowski's work from the 1920s in a positive sense, and seems to have completely ignored his writings after 1930, demonstrates the depth of their personal rupture.[7] He never seems to have directly addressed the fact that the man who had been one of his closest friends and supporters in the 1920s had completely turned against the political views that he was increasingly embracing.

By this time, Pitt-Rivers had undergone another change as well. After inheriting his estate in 1927, he had taken an immediate interest in agricultural matters and farming policy. This was not a wholly unexpected direction for him to take: as the owner of a vast estate, he had a strong vested interest in agricultural matters and as a landlord he had a traditional responsibility to protect the interests of his tenants and champion their views. One of the foremost political disputes facing rural Britain in the early 1930s was the question of tithes, and as a result of his position as a landowner, it was on this issue that Pitt-Rivers would enter the political sphere.

The roots of Britain's tithe system dated back to the early medieval period and were mired in incredible complexity. Under the original version of the system, every farmer and agricultural producer was obligated to give 10 per cent of their products to the church on an annual basis, while the landowner had an obligation to pay for repairs to the local church, as needed (a tradition called chancel repair). This was an important means of supporting the church in the Middle Ages, but by the nineteenth century it had become impractical to require landowners who possessed far more than a subsistence farm to turn over huge amounts of actual harvested goods, and as a result, the Tithe Act of 1836 had converted these obligations to money rather than the delivery of actual products.[8] A vast survey was subsequently made of the country's tithe districts, listing the expected value of crops and enumerating what the residents would be expected to pay, essentially saving the system by tying the tithes to the amount and the quality of the land one owned.[9]

This already-complicated situation was made worse by the details of collecting payment. In 1891, a law was passed making it illegal for a landowner to pass the cost of their tithe obligations on to the tenants renting from them. This made landowners responsible for all of the tithes owed by their renters, but in practice many landowners simply passed the charges along to their renters regardless.[10] In the event that a tithe was not paid it was considered chargeable on the land, meaning that if the property were later sold, the new owner would be responsible for the unpaid money. Even more complicated circumstances came into play if a landowner sought to break their land up into many parts (into a housing development, for instance). Under the law, the tithe owner could then request the entire value of the tithe from the owner of any part of the land. In essence, a person buying a small part of an estate could in theory be asked to pay the entire outstanding tithe for the entire former property, with difficult legal procedures as the only possible remedy.[11] Mercifully, there was a statute of limitations: no more than two years of unpaid tithes could be collected through the courts, and for ecclesiastical tithes, claims could no longer be brought after two ministers had held the local benefice plus six years after the appointment of a third, at a minimum of sixty years, or, as it was popularly known, 'two parsons and a bit'.[12] If an unexpired tithe was not paid, the church could conceivably seize the property from the offending landowner using its own bailiffs.

By the 1930s, the extreme complexities and bureaucracies associated with maintaining this system in the modern world had become overwhelming for everyone involved. Britain was no longer a nation of large landowners and their tenants, and the fact that tithes could persist even after the sale or inheritance of land was fraught with legal complexities. In addition, Britain was no longer a religiously monolithic country: by the twentieth century there were many landowners who were not Anglican and therefore objected to supporting a church to which they did not belong. Tithes had already become controversial elsewhere outside England, and there had been outbreaks of violence over them in Ireland and Wales. The idea of a compulsory payment to the established church with which increasing numbers of residents were no longer affiliated struck many as an affront.[13] British agriculture itself was in a serious slump by the early 1930s, making the tithes even more onerous on those forced to pay them. Most Conservative politicians of the era believed in upholding or increasing protective tariffs on foreign products to help British farmers by maintaining high market prices, but this naturally had the effect of raising food prices for people living in urban areas and was not seen as a long-term politically viable solution. To make matters worse for everyone, the average British agricultural worker made only about a third the wage of the average industrial worker, and agricultural workers were not covered by National Insurance until 1936.[14]

Given all this, there was little doubt that the farmers had a legitimate complaint: after all, tithes were paid only by those engaged in agricultural professions, not by the urban classes that had long since supplanted most rural farmers and landowners in wealth. Pitt-Rivers' own sceptical views towards religion and the church made him even more offended by the fact that he was obligated to support an institution he personally opposed. Tithes were, he claimed, an 'unjust tax' and 'an important factor in raising the production cost of home-grown food'.[15] In addition, tithes were 'excessive by 100 per cent' and 'farms were being left derelict because of it'.[16]

By the late 1920s, Pitt-Rivers had decided that the only way to make progress on tithes was through direct political action. The organization he would use to advance the issue would be one of his own devising called the Wessex Agricultural Defence Association (WADA). Critics smirked that the name itself made no sense, let alone the body's objectives: Wessex was one of the ancient Anglo-Saxon kingdoms, had ceased to exist in the tenth century and was 'not a term known in law at the present time', though it had appeared in Thomas Hardy's novels about the area.[17] Pitt-Rivers insisted the name was accurate because the organization encompassed much of the ancient kingdom of Wessex and had branches in Dorset, Wiltshire, Somerset and Hampshire. He enjoyed the grandiose title of Chairman, giving him latitude to negotiate on behalf of the organization – a power that would soon become important to its political direction – and publish in its official voice. 'Few people can discuss the tithe question more learnedly [than Pitt-Rivers]', the *Dorset County Chronicle* gushed, describing him as a 'wealthy man defending poor men when they are attacked by officialdom' in 1937.[18] While remaining numerically small, the WADA enjoyed at least some support from local farmers seeking political representation on rural issues.[19] While ostensibly based in the nearby town of Sturminster Newton, it is clear that much of the WADA's activity was directed from Pitt-Rivers' own estate. It would soon become obvious that his ambitions for the organization extended beyond political representation for voiceless farmers.

The addition of these political involvements to his already-full schedule created a great deal more work for Pitt-Rivers, who was occupied with not only estate management but also his continuing scientific activities. While he already had a small staff working at his estate primarily dealing with the day-to-day running of its agricultural and financial aspects, he did not have a permanent personal assistant to help with his own affairs. He had already attempted to hire a number of attractive young London secretaries, but all quickly left Hinton St Mary when they began to suspect that their employer had ulterior motives and desired more than just their typing and filing skills.[20] In August 1934, however, Pitt-Rivers hired a young woman who would not only agree to stay but soon become more than just his assistant.

Catherine Dorothea Sharpe had been born in Birmingham in 1914, making her more than two decades younger than Pitt-Rivers and only a few years older than his eldest son. He must have noticed that she had been born the same year he was wounded in the First World War. Sharpe was hardly the average London secretary. Her father was a churchman and moved the family to South Africa to become Archdeacon of Basutoland (modern-day Lesotho) and Director of Native Missions there. The Sharpe family later moved to Windhoek (now in Namibia), where the local Afrikaner, English and German communities were often in conflict with one another.[21] Following a family row over her befriending of a German parson, Sharpe decided she had to escape her mother's overbearing influence and moved first to Johannesburg, where she took on several odd jobs, and later to England, where she lived in London before enrolling at Bristol University with financial help from her grandmother.[22] Her passion was philosophy, and she read psychology as a minor subject and attended lectures on French and German literature.[23] In the course of her studies, she spent a year studying in France and visited a German friend in Berlin shortly after the seizure of power by the Nazis.[24] She, like Pitt-Rivers' wife, was also distantly related to the Mitford family and would later meet Hitler on the introduction of her cousin, Unity Mitford.[25]

Sharpe was intelligent and well read, and her academic interests corresponded remarkably well with those of her employer. A leggy brunette, she possessed a seductive quality that was widely noticed by the men she encountered and was even commented upon by the Security Service.[26] Pitt-Rivers would later refer to her as a 'Jezebel', and it was no coincidence that he nicknamed her 'Becky' in a clear reference to the seductive social climber Becky Sharp in William Makepeace Thackeray's *Vanity Fair* and the subject of a popular 1935 film adaptation. Sharpe herself claimed to 'never discover that there was any similarity' between herself and her namesake and found life at Hinton to be a 'strange combination of secretarial work at an intellectual level and such mundane things as helping arrange estate matters'.[27] Her free time was spent wandering the estate, taking shots at rabbits in the hedges with a rifle and learning how to milk cows.[28]

Sharpe found her employer 'slight, good-looking and imperious, sporting a typical guardsman's moustache, priding himself upon his refusal ever to indulge in cant or humbug. To say he was often brutally candid was no exaggeration.'[29] She later recalled playfully teasing him after he had recalled her from a holiday to help him prepare a paper for a conference in Berlin:

He apologized, with a smile, for breaking into my holiday.

'You can have a full week as soon as we get back from Germany' he said in a placatory tone. I came smartly to attention and saluted.

'Ay, Ay, Sir', I said, looking him straight in the eye.

'Lord, you are a comic girl. You don't have to address me as though I were a naval officer you know.'

'It's your big bristly moustache that does it', I said. [']You're a Royal Dragoon, aren't you? You ought to be carrying a little military cane about with you to whip your secretary into line.' The urge to tease him was irresistible.

I could see that just for a moment he was uncertain whether I was joking or not, but he was seldom pompous.

'It's your damn colonial upbringing that makes you so recalcitrant,' he said with a grin. 'Now be a good girl and fetch your note book and we'll sketch this thing out.'[30]

By late 1934 she had become his lover, and by March 1935 she had moved to sleep in a bedroom directly adjoining his.[31] Three months later, Rosalind told her husband that their marriage had become a 'farce' but she would not leave unless he gave her 'another house and a sufficient income'.[32] If he was unwilling to do so, she said, they would have to find a living arrangement in which they would remain married but she would have complete independence. Unwilling to suffer the financial consequences of another divorce, Pitt-Rivers would soon be taking both his wife and his mistress to international scientific conferences and political events.[33]

In the meantime, however, Pitt-Rivers now had political ambitions based in the tithe issue. One of his first approaches was to newspaper magnate Lord Beaverbrook, who had begun a campaign in 1930 to protect British agricultural interests through the imposition of tariffs on imported food.[34] Pitt-Rivers offered the use of his estate and resources by Beaverbrook's Empire Free Trade Crusade in return for his support on the tithe issue, but Beaverbrook responded that he was 'not an expert on this very complicated problem' and would not be drawn into the campaign, in part because 'I come from a country [Canada] where tithes do not exist...I feel that it is better for me to confine myself to the single broad issue of agricultural protection.'[35]

Chastened by Beaverbrook's rejection, Pitt-Rivers soon sought out friendlier allies on the far right where he had already cultivated connections. In July 1934, he invited British Union of Fascists (BUF) propaganda leader and eventual Nazi propagandist William Joyce, later known as 'Lord Haw Haw', to speak at an event, bizarrely on the topic of tribal government among Polynesian Maoris.[36] The topic was decided because Pitt-Rivers had already written his own apolitical speech for the venue and Joyce demanded fifteen minutes to address the crowd as 'an exponent of the Corporate State movement' and presumably extol the virtues of fascism. The resulting speech on the Maori was evidently terrible, as Pitt-Rivers himself later admitted.[37] Privately, Pitt-Rivers conceded that his views on agriculture were coloured by his views towards the Jews, telling an associate in 1936 that 'It is only the Jews and financial entrepreneurs who profit by war. They have imposed their economy on England and our movement is a movement of producers to free themselves from the interests of parasitic interests.'[38] It was increasingly clear to any informed observer that by the mid-1930s, Pitt-Rivers and the WADA were cultivating close connections with the fascist movement.[39]

In the autumn of 1934, BUF leader Oswald Mosley himself visited and spoke to a Dorset crowd at Pitt-Rivers' invitation. By background, Mosley was a First World War veteran who first entered politics as a Conservative and then crossed the floor in the midst of the controversy over the use of the paramilitary Black and Tans in Ireland. He then joined the Labour Party but resigned from Ramsay MacDonald's government when it failed to adopt his interventionist economic ideas to combat unemployment during the Great Depression. In 1931, he formed a political organization called the New Party to champion his ideas but failed to win any seats in the year's general election and subsequently began to study the political changes taking place in Mussolini's Italy. In 1932 he established the BUF, a movement described by a sympathetic writer as 'a movement freely acknowledging its debt to Hitler and Mussolini, but, nevertheless, distinctively British in policy and method, concerned solely with the welfare and greatness of the British people'.[40] Mosley's followers, called the Blackshirts thanks to their distinctive uniforms, were soon involved in violence against opponents and Jews in the streets of London, most famously at a Mosley-led mass rally that turned into a bloody brawl in 1934 and the violent Battle of Cable Street in 1936. However, at its peak the BUF claimed to have 50,000 members, making it one of the country's largest political parties by membership at the time.[41] Mosley would later rename it the British Union of Fascists and National Socialists, reflecting its German as well as Italian influences.[42]

Thanks in part to the support of *The Daily Mail,* a populist newspaper, most of the BUF's membership was concentrated in London and other urban areas but there were rural chapters as well.[43] Mosley was acutely aware of the political opportunity the unpopular tithe system might present, and in 1933, he and the BUF leadership decided to expand the party's appeal in the countryside by vocally taking the side of the farmers. That summer, Blackshirts set up camp near Wortham, Suffolk, on the land of a farmer who refused to pay tithes to the church. Hoping for an armed battle with the church's representatives, the Blackshirts dug trenches and fortified the farm. This political theatre was brought to an end when fifty London policemen arrived and arrested nearly two dozen of the fascists who later pleaded guilty to the resulting charges. The violent confrontation with the church they had hoped for never materialized. At the same time, though, Mosley had seemingly made himself the champion of the farmers fighting what they viewed as an unfair and oppressive system. One BUF member who farmed land in East Anglia claimed that the tithe system had become so oppressive that some farmers had even committed suicide with their own guns 'because of the mercilessness of the Church and the tithes'.[44]

This view of Mosley as the champion of rural interests and an opponent of the church drew Pitt-Rivers towards the BUF, though he resisted actually joining the party. With the 1935 general election approaching, however, he believed that it was time to make a political stand on the tithe issue and the Blackshirts were the natural allies for such a fight. The first step was the creation of a new political organization separate from the WADA, called the North Dorset Agricultural Defence League (NDADL), for the explicit purpose of supporting a future parliamentary candidate already identified as Pitt-Rivers in its earliest documents.[45] The political platform of the NDADL was barebones at best, with only five ambiguous planks including 'We want something done!', 'Employment before doles!' and 'Unite for rural prosperity!'. On more specific policy questions, it simply stated that it 'supports the agricultural policy of the Wessex Agricultural Defence Association'.[46] Rather than an actual political party, the NDADL was merely a vehicle to support Pitt-Rivers' future candidacy.

Simultaneously, a West Dorset equivalent organization was created to support a candidate running in that constituency.[47]

In February 1935, Pitt-Rivers convened a secret meeting at Hinton St Mary. Attending was the proposed West Dorset candidate, Ronald Farquharson, and two political organizers from the BUF. The fascist representatives agreed to provide behind-the-scenes support for Farquharson's candidacy by paying the salary of his political organizer, who was also serving as general secretary in the WADA.[48] However, Farquharson wisely soon abandoned his plans to run in West Dorset, leaving the future of the agreement uncertain and putting BUF support for Pitt-Rivers' own future parliamentary campaign in jeopardy. Writing to Mosley directly, Pitt-Rivers tried to salvage BUF support for his campaign, telling him that, 'I know it is unnecessary for me to assure you of my belief in the cause for which you stand. I firmly believe that you in your wider movement and I in the agricultural world and particularly in this County are fighting for the same principles and for the same end.'[49] This was a letter that Pitt-Rivers would later regret.

Pitt-Rivers and the BUF leadership soon negotiated a settlement: the WADA political organizer for West Dorset would have his salary paid off and his position terminated as there would be no campaign there. In his place, a Blackshirt political organizer from London would prepare to run a declared fascist candidate there.[50] In return, Mosley himself agreed to a backroom alliance with Pitt-Rivers. Their agreement had five points:

1. The BUF will concentrate immediately on the organizations of the West Dorset constituency, and all Fascists within reach of this constituency are instructed to take part in this work.
2. The BUF will not, in any case, contest every seat in Dorset at the next election.
3. In view of this fact, the BUF will not contest North Dorset at the next election.
4. Capt. Pitt-Rivers and his organization are entirely independent of Blackshirt organization.
5. Members of the BUF will therefore not be instructed to support the candidature of Capt. Pitt-Rivers, but members of the BUF, resident in the North Dorset constituency will be free to assist his candidature in view of the fact that his agricultural policy coincides very closely with the policy of the Fascist movement and his return would be an advantage to the interests of agriculture.[51]

This agreement was obviously intended to establish a climate of mutual support between Pitt-Rivers and the BUF without associating them directly. Pitt-Rivers was not an actual member of the party, and despite his stated support for Mosley's political platform closely guarded his political independence. However, fascist support for his campaign was essential. Pitt-Rivers himself was quick to promote the fact that while he supported aspects of Mosley's platform, he was not in fact a BUF member and stood for agricultural interests alone.[52] In April 1935, Beaverbrook asked Pitt-Rivers directly whether he had 'any leaning in the direction of Fascism', to which he gave a convoluted reply reiterating his independence from the BUF but predicting that 'the Blackshirt organization will announce that…they will not put up a candidate against me, and will leave their own members free to support me should they wish to do so'.[53] This was, of course, more than a mere prediction because he had already engineered precisely this agreement with Mosley.

In October, Prime Minister Stanley Baldwin called an election and Britain's political parties began to campaign. Despite his past bravado, however, Mosley now announced that the BUF would not field candidates, including himself, and would instead campaign for voters to abstain from voting with the slogan 'Fascism Next Time'.[54] This was probably a deathblow for Pitt-Rivers' candidature, if he had any hope at all. Running as an 'Independent Agriculturalist' candidate, Pitt-Rivers was a fiery and aggressive orator, claiming that only by electing an independent candidate could the residents of North Dorset have their views represented because all the major parties were dominated by urban voters.[55] Using the slogan 'peace and the plough', he denounced all other political parties including the 'capitalist Labour Party' and his supporters suggested conspiratorially that the Conservative incumbent would benefit financially from a drop in agricultural prices because his 'commercial interests' would lead him to buy products cheaply and profit handsomely from them later.[56]

The campaign did not go well. Pitt-Rivers often faced hostile crowds of farmers who were sceptical of his actual agricultural knowledge, and his aggressive rhetorical style endeared him to few. Seeking to establish a new narrative, Pitt-Rivers' press officer adopted a new strategy: inviting London reporters to attend a campaign event, he would instead detour them to a local pub, buy them a series of drinks, announce that they had sadly missed the event and hand them a pre-written account of what had supposedly taken place there.[57] When the polling day came, the result was predictable. Pitt-Rivers polled a mere 1,771 votes against the winning Tory candidate's 13,000 and the second place Liberal's 9,800.[58] With this poor performance, he lost his election deposit but was proud to have beaten the Labour candidate in the race by a mere 400 votes. In a twist of convoluted logic, he viewed the crushing defeat as a moral victory: the Conservative voters had cast their papers to keep out the Liberals, he believed, while Liberal voters had done the same against the Conservatives. Only voters casting their ballots for him and the Labour candidate were actually voting for an affirmative position, and, since he had beaten Labour, he had actually won.[59]

Nationally, the North Dorset vote was largely replicated: the Conservatives lost eighty-three seats but still held a majority of 255 in the House of Commons. Despite years of political campaigning, Mosley's

BUF had made no national electoral impact and Pitt-Rivers' candidacy had been a complete bust.[60] Logical contortions aside, as it turned out Pitt-Rivers might well have had a greater political impact by sitting the 1935 election out. In 1937, the Tory MP who won the race died, sparking a by-election. Pitt-Rivers sat this campaign out, but the ensuing election between a Conservative and a Liberal candidate was decided by a mere 543 votes. Had he entered the race, Pitt-Rivers might well have been the spoiler that determined the result.

The tithe issue itself soon reached parliament but its outcome was not what Pitt-Rivers had hoped. Realizing the political extremism the issue was creating in rural areas, parliament reformed the system in 1936, introducing the gradual extinguishing of tithes over a sixty-year period. The reform was an outrage to Pitt-Rivers, who signed a letter denouncing it as a 'betrayal' and helped organize a 5,000-person march in London to protest the bill's continuance of the system and ask the King to withhold Royal Assent.[61] The protest had little effect, and the tithe system remained in place until the Finance Act 1977 terminated the collection of tithes several decades before their formal end in 1996.[62]

His foray into electoral politics thus ignominiously ended, Pitt-Rivers now returned his attention back to his scientific pursuits. The political venture had been a costly failure, and he admitted to Reginald Ruggles Gates that it had distracted him from his academic work.[63] He did not have long to wait before returning to the fray, however. In April 1934, the leaders of the RAI and the Institute of Sociology had convened a special committee to critically consider and 'clarify' the definition of race and its relationship to cultural development.[64] Its originator was C.G. Seligman, who Pitt-Rivers had revered since the early 1920s, and Raymond Firth served as the committee secretary. The Race and Culture Committee, as it became known, included many of Britain's leading scientific figures, including anatomist Grafton Elliot Smith (as committee chairman), left-wing biologist J.B.S. Haldane and anthropologist Geoffrey Morant. As a testimony to the continued respect he enjoyed in anthropology circles, Pitt-Rivers received an invitation to join, as did his friend Gates.[65]

While the committee was theoretically interested in only examining a single issue – 'the significance of the racial factor in cultural development' – Seligman hoped to use it for a more direct purpose.[66] He soon made it clear, behind the scenes, to Firth and others that he hoped the committee would scientifically attack Nazi concepts of both a distinct Aryan race and the Jewish 'race' in a report that would be accessible to the general public. He made this intention clear in a letter to Elliot Smith:

> I think the Committee should consider the preparation of a relatively short – say 15 or 20 pages – statement, which could be put before a general public. My reason for bringing this forward now and considering it urgent is the altered state of things in Germany [following

the Night of Long Knives purge in 1934] .... Although we cannot influence Germany, I think it would be well to have a statement before the public in order to take whatever chance there may be of influencing public opinion against the continuance of the more extreme results of the present Aryan fallacy when a new regime emerges....Prophecy is of course a dangerous game, and you may think I have been foolish to indulge in it, but even if there is no change in Germany I still think some such statement as I have suggested might be very usefully prepared during the summer holiday.[67]

Anticipating German criticisms of the report, Seligman, who was himself Jewish, opted to take no public role in the committee, and a number of members later rejected the inclusion of London School of Economics sociologist Morris Ginsberg to the committee on anti-Semitic grounds.[68] Ginsberg eventually took a role as an un-credited 'adviser' to the committee, and Seligman pushed for the creation of a 'non-Jewish travel fund' to provide funds for committee members to attend meetings.[69] By thus excluding Jews from the official membership of the committee, Seligman hoped to produce a scientific document that was pre-emptively immune from the allegation of 'Jewish influence'.

Following a few meetings and with the sense of urgency that Seligman had encouraged, geographer H.J. Fleure and Firth sent out a draft statement to all members in February 1935 stating that the committee had agreed that race could be scientifically defined as 'a number of persons possessing in common a number of innate physical characteristics, and, by interbreeding, normally transmitting these to their descendants', while a cultural group was defined as 'a number of persons possessing in common and normally transmitting by educational means to their successors a number of distinctive social activities'.[70] Pitt-Rivers was outraged by these definitions. His own anthropological research was predicated on rigid notions of race derived from Arthur Keith's ideas about reproductive and evolutionary isolation, not that idea that a race was simply a group that bred together. In addition, his belief in biological 'culture-potential' indicated that races were not intrinsically equal but hierarchical, and the committee's definitions seemed to discount this aspect in favour of the implicit view that all races might be equal. Finally, he quickly detected that the committee's intentions towards German notions of the Aryan and Jewish races might be hostile.

In an effort to pre-emptively head off the committee members who might be inclined to endorse a statement attacking Germany, Pitt-Rivers privately made overtures to his contacts in the country to help him present the Nazi response. In April 1935, he wrote to Königsberg racial anthropologist Lothar Loeffler asking for the German perspective on the committee's discussions. Pitt-Rivers knew Loeffler, a hard-core Nazi racial hygienist and anti-Semite, through his connection with his IFEO colleague

Eugen Fischer.[71] Loeffler's views on race were well known in Germany and he had built a reputation for towing the Nazi party line by refusing a chair at the University of Frankfurt out of a stated contempt for the area's Jewish population.[72] Seeking help against the emerging majority on the Race and Culture Committee, Pitt-Rivers passed Loeffler copies of key committee documents along with the contrarian statement he hoped to include, asking if he had 'correctly interpreted the attitude of German anthropologists'.[73]

Loeffler was happy to oblige Pitt-Rivers' request. In a scathing memorandum provocatively entitled 'Scientific as Against Political Implications of the Aryan Question', Pitt-Rivers summarized Loeffler's views to his fellow committee members, taking care not to reveal the identity of his German contact:

> Owing to the prominence that has been given to the so-called *Aryan* controversy and the use of the word *Arier* in the Aryan clause of the Nazi Government policy, discriminating between the Jewish and non-Jewish [*sic*] German populations, the following note is added as an explanation of the attitude of German Anthropologists.
>
> It is felt that in Germany as in England, popular and political use of anthropological terms such as *Aryan race, Semitic race, English race, Celtic race,* etc. should not be interpreted as the adoption in either country of a scientific terminology by scientific anthropologists.
>
> The following statement emanates from an academic source and is communicated *privately.*
>
> The German Government privately agree that the public use by them of the word *Aryan* is unscientific, but it has now acquired a new meaning as a result of their use of it, and to change the word would be interpreted as dropping the *Aryan* clause – or weakening it .... German University Anthropologists interpret their duty as to back up their Government politically but to secure their scientific position in a way that does not interfere with that. The original *Aryan* clause was issued in a hurry without consulting the Scientists.[74]

This was a clever bit of sophistry. By declining to scientifically defend the concept of the Aryan race, Pitt-Rivers and Loeffler were effectively arguing that any criticism of the idea was intrinsically political. As a result, it would be interpreted as an attack on the German government's policies rather than the pronouncements of the country's scientists. They had also astutely sidestepped the question of whether the very idea of the Jewish race was scientific or political, simply suggesting that any discussion of race was beyond the scientific sphere entirely. Finally, the fact that Pitt-Rivers had not named his source of this information made it un-attributable and suggested that it could have come from a government official even more senior than Loeffler.

Any remaining comity on the Race and Culture Committee was brought to an end with this bald attack on its very existence. With Pitt-Rivers and other contributors refusing to budge from their individual positions, the idea of producing a single statement laying out the scientific definition of a race was abandoned in favour of a short pamphlet that began with an opening consensus section that included two definitions of race, followed by a series of dissenting statements by individual members. The first definition appears to have been agreed on by all contributors but Pitt-Rivers and Gates and cast doubt on the scientific ability to define race as a meaningful concept:

> A Race is composed of one or more interbreeding groups of individuals and their descendants, possessing in common a number of innate characteristics which distinguish them from other groups....This definition may in some cases apply to the whole population of a particular area if it breeds freely; it may also apply to an interbreeding portion of a group within a particular area provided that this portion carries a number of common innate characteristics distinguishing it from other groups.[75]

The competing definition, written by Pitt-Rivers himself, was simpler:

> By Race is meant a biological group or stock possessing in common an undetermined number of associated genetical characteristics by which it can be distinguished from other groups, and by which its descendants will be distinguished under conditions of continuous isolation (*i.e.* so long as the stock is preserved against internal dilution).[76]

The differences were clear. While the first definition left open the possibility that a race was simply another term for the population of a physical place, freely interbreeding or breeding only within an internal group, the second argued that races required isolation and protection against interbreeding, in line with Keith's long-standing views. This was the view of native races in the South Pacific that Pitt-Rivers had argued for in *The Clash of Culture*, and this was now the view that he believed proved the need to protect the Aryan race against racial dilution.

Even the inclusion of this definition was not enough, however, and Pitt-Rivers insisted he be allowed to include an individual statement in the Race and Culture Committee's report dissenting further from his colleagues. In it, he reiterated his argument for the validity of German racial views on the grounds that they were no more unscientific than any other view:

> No doubt race-culture problems, in their strictly scientific bearings, inevitably attract the intrusion of political interests which, however legitimate in their own sphere, cannot but serve to obscure anthropological solutions and to promote national and racial antagonisms. These

obtrusive interests are conspicuous in many countries, and the term Aryan has been used in Germany to focus political attention on the Jewish problem. *Aryan race* no less than *English race* should not be interpreted in either country as the adoption of a scientific terminology.[77]

Pitt-Rivers' efforts did not end here. With the committee finally about to publish a short pamphlet in November 1935, he abruptly cabled Firth to accuse his fellow committee members of 'mutilating' his statement and demanding the ability to modify the proofs at the final moment though he had evidently orally agreed to them in a meeting.[78] Firth was understandably outraged by the demand but capitulated, and Pitt-Rivers both modified his own contribution and added two critical footnotes to the consensus section of the report. The second of these was the insertion of a reference to *The Clash of Culture*, making it the only act of self-citation by any committee member in the opening section of the report.[79] While other authors mentioned their past work in their individual statements, to insist on the citation of his own work was an act of extreme arrogance, as was the fact that he had done so in a last-minute manner that had prevented other members from objecting, lest the report be further delayed.[80]

When it finally emerged in print in 1936, the Race and Culture Committee's report was an obvious defeat for Seligman, Firth and other anti-racist voices in the British anthropological establishment.[81] Pitt-Rivers had nearly single-handedly been able to subvert Seligman's intentions, making it a convoluted jumble of ideas and contradictions rather than the brief anti-Nazi statement the body had been set up to produce. Both Pitt-Rivers and Gates saw this as a major victory for their scientific views and Anglo-German relations, and the former subsequently assured German racial hygienist Karl Astel that the two men were 'working for a better understanding with German men of science'.[82] Yet, this victory had come at a hefty price: Seligman had been one of Pitt-Rivers' earliest academic supporters, had facilitated his connection to Malinowski and proposed him for membership in the RAI. Firth had, in some senses, been one of Pitt-Rivers' intellectual heirs, citing *The Clash of Culture* in his doctoral thesis and having Pitt-Rivers as his doctoral examiner. It was no secret what Pitt-Rivers had done to the derail the committee, and while Firth remained a social connection their relationship had clearly suffered a blow.[83] Sending Pitt-Rivers a copy of the final committee pamphlet, Firth facetiously played upon his anti-Semitism and warned him not to 'get Judomania!'[84] Humour aside, the academic world was beginning to close its final doors to Pitt-Rivers in the face of his increasing extremism.

It was also clear to many observers that Pitt-Rivers was almost completely infatuated by Hitler and Nazism by the mid-1930s. He had cleverly couched his opposition to the Race and Culture Committee's report in a scientific guise, but he had privately admitted to Astel that his efforts had been driven by an interest in protecting the German regime's reputation. He had already made several visits to Germany since Hitler's seizure of power, telling Gates in 1933 that during German university lectures his definition of race, identical to that expressed in the Race and Culture pamphlet, 'appeared to get general endorsement'.[85] In 1935, he dispatched his eldest son Michael to Munich to study anthropology in preparation for his future studies at Oxford.[86]

Some in Pitt-Rivers' circle now began to express their concerns. Despite his own past flirtations with Mussolini, Oscar Levy was increasingly worried that fascism would not live up to his Nietzschean aristocratic expectations. By 1935, he was denouncing the 'leftist' and populist tendencies of both Mussolini and Hitler, warning Pitt-Rivers about their corrupting and destructive potential. 'I likewise see that you are now taking interest in politics and that you expect much from the new Fascistic or Nationalistic movements', he wrote. 'So did I once: you remember I was the first to call upon Mussolini and I am responsible for the German connection with him. I regret it now. I now think that Fascism as well as Bolschevism [sic] are movements of the Left and that the new message will only be carried through against both these popular movements.'[87] Levy's hopes that fascism would establish a new 'aristocracy' in the way he imagined had been disappointed by their popular appeal.

At the same time, many of Pitt-Rivers' supporters still stood by him even as his own views became increasingly extreme. Despite his relationship with Malinowski breaking down, he still enjoyed the important friendship of both Gates and Arthur Keith. In late 1935, both men co-sponsored his membership in the Athenaeum Club, one of London's most exclusive gentlemen's clubs. Charles Darwin, Francis Galton and his grandfather, General Pitt-Rivers, had all been members, making it the ultimate social club for him to join.[88] Keith was essentially his intellectual grandfather on the concept of race, and both Pitt-Rivers and Gates referred to him reverentially at every possible turn. Still enjoying the status of a revered scientist, Keith's approval of both Pitt-Rivers' and Gates' work gave them a powerful ally in the years before the Second World War. Pitt-Rivers even evoked Galton himself to express his support for Keith's rigid view of racial difference, condemning the Eugenics Society in the process:

> Galton's values were those of the 'endemic' and national, in which sense, if so foolish a phrase must be used, he was 'fascist'. He was a great methodologist, he thought like an artist as well as a man of science, and he and his class, intellectual as well as social, are almost extinct. He would certainly have agreed with, *inter alia*, Sir Arthur Keith on the eugenic value of 'racial prejudice' against the current popular 'egalitarian and democratic prejudice', on the question of the inheritance of mental abilities and psychological racial characteristics his views were fundamentally opposed to modern democratic sentiment, which the Eugenics Society seems to concerned to appease.... In pursuit of my

'unpopular mission', allow me in temerity to state without equivocation that what is now advocated by 'our' Society…. References to Racial betterment or welfare and to cultural progress in pamphlets and speeches recur as frequently as references to 'race prejudice', but great care is taken to avoid any definition of either 'race' or 'culture'.[89]

Supported by Keith and Gates, Pitt-Rivers was thus undeterred in his pursuits. In 1935, the IUSIPP, of which he was still general secretary, held its general assembly and international conference in Berlin. The event had originally been scheduled to take place the previous year, but the American delegation was concerned about the political implications of holding an event in the country while the Nazis were in the midst of purging Jews from the country's universities. Ignoring the objections, the event went forward anyway, with Pitt-Rivers travelling with Sharpe to Berlin to make many of the arrangements directly with Fischer and his other contacts.[90] When it finally took place, the conference was a disaster for the IUSIPP. Nazi officials and a parade of disreputable scientists used the platform to pontificate on the merits of the most racialist varieties of eugenics. The fact that the venue was full of leading scientists from around the world gave their statements instant credibility and granted a veneer of legitimacy to fundamentally unscientific proclamations. In his opening statement, Sir Charles Close attempted to state that the event was purely scientific rather than political, but no one seems to have been fooled by this declaration. The published proceedings of the event included dozens of papers praising the racial policies being carried out by the regime, essentially handing the German government a ready-made piece of propaganda.[91] The idea that the Berlin congress could have ever been apolitical was a dangerous delusion and through his own involvement in the planning Pitt-Rivers had given the Germans another gift, just as he had on the Race and Culture Committee.

By mid-1936, Pitt-Rivers' personal life was in nearly complete disarray. Still married to Rosalind, his relationship with Becky Sharpe was becoming socially complicated. In June, he proposed going on a holiday to Cornwall with her but feared the 'conventional difficulty of going away together'.[92] The following month, he and his wife exchanged words in his bedroom while Sharpe sat nearby. After a subsequent row between them, Rosalind begged Becky not to leave, telling her 'he needs you far more than he needs me. I couldn't remain in the house alone'.[93] Still unwilling to grant his wife a divorce, Pitt-Rivers was stuck in an increasingly untenable personal situation. By August, he could tolerate it no longer and left England to embark on a five-month visit to Europe with Sharpe. Arriving on the spectacular Bavarian island of Herreninsel, he now set to work on a second edition of *The Clash of Culture*, using her as his personal assistant.[94]

Visiting Munich on his travels, he was outraged to find Michael 'wasting his time away with a lot of young English friends' rather than learning about National Socialism and promptly sent him to the estate of racial hygienist Alfred Ploetz, where he hoped he would be instilled with a greater understanding of both science and Nazism.[95] *Mein Kampf* was assigned reading for the visit, but Michael was unimpressed with the book, telling his father in no uncertain terms:

> I was, on the whole, rather disappointed. Though I agree that the author's courage and deep feeling are impressive, I think it scarcely fulfills the task of explaining the aims of his movement and of drawing a picture of how it developed, for according to the preface it was written for the sake of those adherents of the movement who belong to it in their hearts and wish for enlightenment regarding it.[96]

Pitt-Rivers' reply to his son was scathing:

> Obviously you have not understood anything at all about 'Mein Kampf'. My advice is to read it through two or three times more, annotating and making notes; and, especially, making notes of what you do not understand because you do not know the facts. Draw comparisons with what you know about English history and politics. Do, my dear Boy, understand that a man who writes of his 'weltanschauung' [world view], which, after twenty years, has become the 'weltanschauung' of a mighty nation, which has challenged the Judaic-Messianic myth of two thousand years, will not be easily intelligible to you. Your superficial comments are *quite* valueless except as showing that you have not begun to understand how to investigate.[97]

For his own part, Pitt-Rivers believed that his study of *Mein Kampf* had revealed the philosophical origins and potential of National Socialism, as he told his friend and Dorset neighbour Rolf Gardiner:

> Whilst here I have been reading 'Mein Kampf', and am particularly interested the beginnings of movements often articulated through the mind of one man, or independently by several men in different parts of the world. The roots of these ideas have their origins in the philosophical varieties of the past. If they articulate the general appetite and need of humanity at some time or other they will conquer, apparently as a new 'weltanschauung', and be ascribed to the man who, by action and fighting when the time is ripe, succeeds in raising the standard which is the rallying point of the new movement. Who that man is does not matter very much.[98]

Gardiner himself would have likely agreed. A blood-and-soil enthusiast, he had renovated an estate he called the Springhead Ring that he used to instil the principles of rural life and traditional English culture on young people who visited in groups. He was anti-Semitic and suspected of harbouring Nazi sympathies of his own, and while he and Pitt-Rivers never appear to have been particularly close friends they certainly shared views.[99]

In 1936, Pitt-Rivers and Gardiner briefly planned to join their forces to protect Dorset agriculture from 'middlemen and Jews' by creating a Wessex alliance similar to the one Pitt-Rivers had imagined during his anti-tithe campaign. Gardiner suggested that Pitt-Rivers should take on the 'scientific' aspects of the campaign while Gardiner would be in charge of the 'spiritual' aspects related to rural renewal and preservation of English folk culture.[100] Throughout his travels in Germany, Pitt-Rivers attempted to gain information about agricultural conditions that he could send back to Gardiner as part of their own reform efforts. Their letters often closed with a line borrowed directly from Nazi propaganda: '*Die Juden sind unser Unglück!*' ('The Jews are our misfortune!').[101] Gardiner likely represented the closest like-minded figure in interwar British agriculture to Pitt-Rivers, and the fact that the two men corresponded throughout the mid-1930s and planned how to take the ideas of National Socialism into the British agricultural world reflects their shared obsession with the tenets of Nazism.

Indeed, with the propaganda coup of the 1936 Berlin Olympics recently concluded, Pitt-Rivers' view of the Nazi regime could hardly be more positive during his visit. Yet international tensions were rapidly growing in a nearby area, and he could hardly resist the temptation to become directly involved himself. By August, the ethnically German population of the Sudetenland in Czechoslovakia was vocally claiming that it was being oppressed by the government in Prague. This complicated situation had been brought about by the 1919 Paris Peace Conference, in which Czechoslovakia had been constituted out of the remains of the Austro-Hungarian Empire. While the Czech-speaking population was a majority of the country, German speakers remained a majority in the Sudetenland border regions. Led by pro-Nazi politician Konrad Henlein, the *Sudetendeutsche Partei* (Sudeten German Party) had won the most seats in the 1935 Czech parliamentary elections and, encouraged by Hitler, demanded increased concessions and regional autonomy from Prague. Stories of oppression and atrocities against the German minority were frequently found in Nazi propaganda and British newspapers throughout the mid-1930s.[102]

Pitt-Rivers had a particular interest in the Sudeten situation: fundamentally, he believed, it was a problem of demography and population. As an anthropologist, he argued that he could bring an important scientific perspective to the situation by conducting fieldwork and help build Anglo-German understanding over the matter. There was a personal element of outrage as well. Pitt-Rivers' scientific contacts in Germany informed him conspiratorially that German-speaking scientists at the University of Prague had been excluded from taking part in the international demography conference he had helped arrange in 1935.[103] Characteristically, he believed these provocations against Germany were part of the same conspiracy he had examined in *The World Significance of the Russian Revolution*. It was all a vast plot, he claimed, by the Jews and the communists to ruin Western civilization: 'From the beginning the Czechs have been used as the tools and the decoy of the Comintern and have been the protected allies of Russian Soviet Dictatorship', he wrote.[104] The fact that the Sudeten issue might lend itself to 'scientific' analysis and contained a personal interference in his own professional sphere led Pitt-Rivers to 'make his own investigations and uncover the truth'.[105]

Naively entering a political hornet's nest under these 'scientific' pretexts, on 31 August, he drove with Sharpe over the German border into Czechoslovakia. Arriving in Prague, the pair spent a few days sightseeing in the capital – she found it to be an 'enchanting city … though the atmosphere was grim' – before driving north-west to the Sudetenland.[106] On 6 September, they were present for a rally of 20,000 ethnic Germans in the town of Meierhöfen led by Henlein and, afterwards, met with him personally over coffee.[107] Sharpe found him to be a 'tall, good-looking square-headed man with a high forehead and horn-rimmed spectacles' who 'smiled constantly at me'. Chatting in French, Henlein told Pitt-Rivers that his party was not interested in an alliance with Germany or unification with the Reich but merely desired Sudeten 'autonomy' and 'deliverance from what the Sudeten Germans in their case considered to be the tyranny of the Czechoslovaks'.[108] These assurances received, Pitt-Rivers and Sharpe drove into the Sudeten countryside to conduct their own first-hand investigations. In rural villages and towns, they saw abandoned factories falling into ruin, high unemployment and inadequate medical facilities. Stopping to interview locals along the way, the pair heard stories of starvation and hardship, recording their progress in dozens of photographs showing shanties, starving children and deteriorating infrastructure.[109] 'It was a gloomy journey and I began to dread each stop we made for these interrogation purposes', Sharpe later recalled.[110]

The hardship and suffering they witnessed was undoubtedly genuine. The Czech economy had been devastated by a deep economic crisis throughout the 1930s, accompanied by a steep decline in exports. Many of the border areas inhabited by the German minority had been dominated by light industry and the manufacture of consumer goods. These industries had been hit particularly hard by the collapse in domestic consumer demand and exports after 1930.[111] By 1936, heavy industry in the Czech interior, including armaments, was well on its way to economic recovery but the German borderlands remained economically devastated. The shuttered factories Sharpe and Pitt-Rivers observed were the result of an uneven economic recovery, not a political plot in Prague.[112] The clearly imbalanced nature of the Czech economic situation had given Henlein and Hitler a grievance they could use as a political weapon to stir up nationalistic resentment against the Czech majority, and Pitt-Rivers would soon adopt their preferred interpretation of the crises.

There was more immediate trouble on the horizon, however. Alarmed by the presence of two foreigners travelling in an expensive motorcar, a schoolmaster in the town of Karlsbad (present-day Karlovy Vary) refused to let the pair look around his classrooms and phoned the police to report the presence of two 'dangerous English spies'.[113] The car was impounded, the camera and film confiscated, their passports taken and, Pitt-Rivers claimed later, 'spat upon...because I happen to be a friend of Konrad Henlein'.[114] At the police station, Pitt-Rivers was 'characteristically aggressive' and blustered pretentiously at the officers while Sharpe sat in silence. She later recalled that it was 'the most terrifying moment of my life...and I feared what would happen to me'.[115] The pair was eventually taken back to their hotel where they remained under guard until the arrival of a British diplomat who secured their release, the return of their belongings and an apology from the police chief for the inconvenience.[116] Wisely leaving for Germany immediately, Sharpe wrote later that she 'felt as though I already knew what a communist prison was like. Czechoslovakia was not under communist control in those days – but a few grim hours in police detention brought all the horrors of such places to one's mind'.[117]

Pitt-Rivers was bombastically convinced that this episode had revealed the depth of the international Bolshevik conspiracy and the lengths his opponents were willing to go in an effort to silence him. His attentions now turned to another front in this purported conflict: Spain. The mid-1936 uprising by right-wing Nationalists against the Republican government had developed into a bloody civil war, with Nationalist troops pushing hard against Republican positions – and communist forces – in the north-west. Apparently learning little about the dangers of wandering through politically sensitive areas, Pitt-Rivers and Sharpe returned to England in October to prepare for a journey to Spain, leaving again almost immediately.

They were now accompanied on their journey by Richard Findlay, a former Royal Air Force reservist who had recently stood for parliament as an 'Independent Conservative' candidate in the 1935 Norwood by-election. He was backed in this quixotic campaign by Randolph Churchill, son of the future prime minister. The younger Churchill had already embarrassed his father by himself running as an independent against a Conservative in a Liverpool by-election earlier in the year. He had intended to break up the governing National Government coalition, but the only effect of his candidature was to allow Labour to win the seat by splitting the vote.[118] Findlay was hardly the ideal candidate to carry Churchill's banner forward. He had resigned from the BUF only days before entering the campaign after purportedly being outraged to find 'evidences of a strong hatred of the Jews – by no means confined to the rank and file – which I cannot support, since I regard it as entirely alien to the British tradition'.[119]

FIGURE 5.1 *Pitt-Rivers and his companion Catherine Sharpe visited Spain shortly after the outbreak of civil war in 1936. By the mid-1930s, Pitt-Rivers' war wound to his left leg necessitated the use of a heavy walking stick.*

His candidature was based on his opposition to 'the shameful fashion in which the leaders of the Conservative Party have perverted the power of the Party to adopt Socialist measures' and he claimed he had joined the BUF only because it was the 'only political organisation that was making any fight against the menace of Socialism'. He now recanted that view after professedly discovering the promise of the younger Churchill's political vision.[120] Despite Churchill's support, Findlay had lost the by-election by a large margin to Duncan Sandys, who would himself soon marry Randolph Churchill's sister. Notwithstanding his stated opposition to anti-Semitism, the first visit made by Sharpe, Pitt-Rivers and Findlay was in Belgium, where they met with far-right Rexist party leader and future collaborationist Leon Degrelle, before journeying on to Paris to meet anti-Semitic author Georges Batault.[121]

From France, the trio drove through the Pyrenees, visiting the resort of Biarritz before crossing the Spanish border and arriving in the Nationalist stronghold of Burgos. Tensions in the region were high, and Sharpe recorded that black-hooded Falange supporters were 'everywhere in evidence' in northern Spain and on their guard.[122]

The trio's visit was carefully stage-managed by their hosts. For three days, Sharpe recalled, they were introduced to 'various dignitaries, some in colourful uniforms – like characters from a Gilbert and Sullivan opera – and others not'.[123] The strategically important Basque town of Irún had been bombarded by the Nationalists only weeks before and subsequently 'reduced to ruins' as the retreating Republicans destroyed anything that might be of value to their enemies. Taken on a visit there by their hosts, the visitors were told that the Republicans had 'butchered hundreds of inmates and hostages' and that the damage caused by the fighting and bombardment 'was trivial with the wholesale destruction by the Reds when they occupied the town'.[124]

Residents in Burgos and elsewhere told them bracing tales of abuse at the hands of communist forces and of secret support being given to the Republican side by the Soviet Union with the help of the officially neutral French government. As evidence, they were shown letters addressed directly to Nationalist leader Francisco Franco and his lieutenants that had allegedly been found in French post-office bags on a captured ship, suggesting that the French had been diverting important Nationalist communiqués to their Republican enemies. Leaving Burgos and driving to the French side of the border with Catalonia, Pitt-Rivers and Sharpe purported to observe trucks loaded with munitions and supplies moving towards Spain without objection from French officials, further suggesting that the French were at least implicitly backing the Republican side.[125]

Returning to England in December, Pitt-Rivers was even more outraged by what he had found in Spain than what he had seen in Czechoslovakia. In Spain, he believed, was a direct war between the forces of international Bolshevism and Western civilization. The French government was secretly backing the world revolution despite its professed neutrality, he claimed, and it was 'generally known' that the British government was sending medical supplies and gas masks to 'the Red Forces alone'. He accused his former Race and Culture Committee colleague, left-wing biologist J.B.S. Haldane, of travelling to Madrid to organize 'chemical gas warfare' against civilians. He now claimed that even the Czechs were involved in supplying 'munitions of war and volunteers to the Spanish Red Forces' through 'Franco-Czech-Soviet military mutual assistance Pacts' that he purported to have been investigating in the Sudetenland.[126]

Convinced that Britain was being dragged into a world conflict on the side of the Communists, Pitt-Rivers took it upon himself to sound the alarm. In a lengthy memorandum to the Earl of Plymouth, the under-secretary of state for foreign affairs and a chairman of the Non-Intervention Committee tasked with preventing foreign interference in the Spanish Civil War, Pitt-Rivers laid out his evidence and accused both the French and British governments of providing covert support to the 'Reds'. The report was clearly taken with an appropriate level of scepticism: Plymouth's private secretary did not bother to even reply for several weeks, and after being pestered by Pitt-Rivers by phone, merely sent a note thanking him for his efforts and assuring him that there was no need to pass along further information about 'sins of omission or commission on either side' that would not 'have any practical effect on decisions as to our course of action'.[127] In a scathing letter to the British ambassador in Berlin, Sir Eric Phipps, Pitt-Rivers was even more direct in his accusations:

The organized massacres and tortures inflicted wholesale on Spanish nationalist hostages, men, women and children, by the Red criminal scum of Europe, which British diplomacy alone acknowledges as the Government of Spain, can only be condoned, encouraged or palliated, and the evidence of them deliberately suppressed, by those I feel humiliated to acknowledge as fellow Englishmen, even when they do not happen to be international Jews.[128]

Writing to the Franco-sympathizing Conservative MP Arnold Wilson, Pitt-Rivers claimed these facts were being systematically supressed in the press: 'Reuter's H.Qr. [headquarters] in Paris instruct their correspondents to avoid any news coming from the White side in Spain [the Nationalists] – anything favourable to the Reds is accepted – any news favourable to Franco is described as "propaganda". I have definite evidence'.[129]

Thus persuaded that the British government was increasingly influenced by the 'international Jews' and was being lured into backing the communist cause, Pitt-Rivers turned to renew his ties with Germany. In late 1936 he began attending meetings of the Anglo-German Fellowship (AGF), an elitist society founded at the behest of German Ambassador Joachim von Ribbentrop to encourage favourable British perspectives and policies towards Nazi Germany.[130] Ribbentrop's initial strategy was to encourage meetings and reconciliation between British and German First World War veterans through moving meetings between the British Legion and other groups with their German counterparts.[131] Pitt-Rivers himself had experienced such a reconciliation at a public event in September 1935 when he met the *Oberbürgermeister* (mayor) of Wurzberg, who, it emerged, had been fighting on the German side at Ypres and had been 'wounded on the same day in the same engagement nearly twenty years ago'. This coincidence was 'acclaimed as a portent cementing the new-found understanding that could never again lead to a war in which Englishmen and Germans would fight on opposite sides'.[132]

Expanding his efforts beyond veterans' groups in early 1935, Ribbentrop was responsible for the creation of both the AGF and its German counterpart, the *Deutsch-Englische Gesellschaft* (German-English Society). Membership in the 'non-political' AGF was deliberately geared towards the wealthy and socially prominent, with luxurious banquets and frequent appearances by Ribbentrop and other prominent dignitaries. The organization produced a slick periodical, *The Anglo-German Review*, that featured ludicrous pieces of propaganda praising virtually every aspect of the 'New Germany' and once describing Hitler as 'not the kind of bogey man the British public, subjected to a steady barrage of alarmist and distorted French Press reports, is led to believe, but a big-hearted man full of human kindness and understanding'.[133] It was a sort of celebrity magazine for the Third Reich, with extensive photo spreads showing Hitler and other

Nazi officials relaxing and spending time with children featured in nearly every issue. A 1936 letter to the editor critiqued the sartorial styling of key Nazi officials, declaring propaganda minister Joseph Goebbels to be 'undoubtedly the best dressed' while Hitler 'is a particularly difficult type of man to dress, and appears also to be quite indifferent to his personal appearance. Clearly he is a visionary – a man of great intellectual driving force, but lacking in the appreciation of material and practical things'. As a result, Hitler 'shines only in comparison with Signor Mussolini, who appears always in my mind a little cheap, sartorially speaking'.[134]

By 1936, the AGF proudly boasted of having more than fifty members of Parliament, several directors of the Bank of England and dozens of other prominent businessmen on its membership rolls.[135] Its chairman was Lord Mount Temple, a former minister of transport, and its honorary secretary was T.P. Conwell-Evans, the former private secretary of a Labour minister and an Oxford graduate who had spent time at the University of Königsberg. Once described as '*the* most ardent Gemanophile in this country', Conwell-Evans helped arrange visits to Germany by prominent Britons including former Prime Minister David Lloyd George, who met Hitler at Berchtesgaden in 1936.[136] He would later recant his pro-German leanings and become an informant for the British government, passing along sensitive intelligence gleaned from personal connections that included his 'bosom friend' Ribbentrop.[137]

The internal workings of the AGF were less glossy and idyllic than the *Anglo-German Review* would have suggested. Pitt-Rivers had been recommended to join the organization by Captain H.W. Luttman-Johnson, a right-wing extremist who also served as the secretary of the January Club, a dining society dedicated to discussions and study of fascism that Pitt-Rivers had joined.[138] Beneath the surface, the AGF was an amalgamation of conflicting agendas: while some members seem to have possessed a genuine desire to improve Anglo-German relations or simply secure favourable business relations between the British Empire and the Reich – British Empire Steel Products Co., Imperial Chemical Industries Ltd. and Unilever all made sizable financial contributions to its operations – there was a contingent of Nazi sympathizers and anti-Semites as well.[139] The organization's paid secretary, Elwin Wright, was a rabid anti-Semite who later openly advocated the mass shooting of Jews and called parliament a 'corrupt body of bastards'.[140] Admiral Sir Barry Domvile, the anti-Semitic former Director of Naval Intelligence and president of the Royal Naval College, joined the AGF along with his wife. Domvile maintained that his only interest was patriotic, and he argued that he hoped to protect the British Empire by preserving Anglo-German relations alongside Britain's naval strength. At the same time, he admitted that he was particularly impressed with Hitler's regime because of its attention to the 'Jewish masonic plot' that he believed was responsible for a vast array of worldwide developments.[141]

The AGF was thus a hotbed of conflicts between pro-business voices trying to maintain Anglo-German relations or simply ingratiate themselves to the German regime and those whose sympathies lay with Hitler and National Socialism. Pitt-Rivers characteristically stumbled into this complicated morass in the most impolitic way possible. In December 1936, a luxurious banquet was held by the organization with Ribbentrop as the guest of honour. Pitt-Rivers and Sharpe were in attendance as prospective members, and a toast was raised to German ambassador. In response, Ernest W.D. Tennant, a British banker and a close friend of Ribbentrop, rose to politely caveat that 'of course we do not all agree with what is going on in Germany' in a clear effort to emphasize the organization's inclusivity. Pitt-Rivers loudly injected from his seat 'Why not?' At an adjoining table the AGF's solicitor, G. Le Blount Kidd, pointed in Pitt-Rivers' direction and remarked, 'if that man applied for membership of the Fellowship I will see that he is not admitted'.[142] Mount Temple blustered to his own table that Pitt-Rivers 'must have either had too much to drink or be suffering from shell-shock'.[143]

In front of Ribbentrop, this was a regrettable incident that was an embarrassment not only to Pitt-Rivers personally but the AGF as a whole. He later claimed that he had 'expressed no desire to join' the AGF before the incident and only attempted to do so at the further request of his friends after it had taken place, which was almost certainly false. Regardless, the AGF's governing council immediately rejected his application for membership while Sharpe's was approved.[144] This obvious insult precipitated an acrimonious series of correspondence in which Pitt-Rivers accused the AGF of doing nothing to counter 'mischievous propaganda' against Germany in the British press while representing only 'commercial interests' and the 'French-Czecho-Soviet "League of Nations"'.[145] Now a member herself, Sharpe appealed against his rejection and signed a letter with six other members demanding an explanation from the council.[146]

While officially the council's deliberations were private, Pitt-Rivers soon found an ally in Wright, who had troubles of his own. While he initially told Sharpe in an official capacity that 'the matter is quite out of my hands' because Pitt-Rivers 'began his acquaintanceship with me by an attack on some who are already members of the Fellowship, and who he thinks are a source of weakness and not of strength', Wright was himself sacked in May and replaced by Conwell-Evans. The impetus for his firing was an incident at the same dinner over the presence of a German national flag bearing the Swastika. Wright had arranged for the flag to be borrowed from the German embassy and placed it next to the Union Jack. Conwell-Evans was outraged by its presence, remarked that 'we shall all be branded as Nazis', and called for it to be removed, but no one could be found who would dare to do so in the presence of the German ambassador.[147] Conwell-Evans subsequently asked Wright whether he was anti-Semitic, to which Wright replied that in his view Hitler's government had 'acted under stern necessity in its treatment of the Jewish minority' and had 'a very strong case for its legislation'.[148] Wright's days were numbered and he soon found himself without employment.

The recriminations for Wright's sacking were immediate. He claimed the action was 'illegal' and the result of a 'political plot to get me out of the way so that complete control of the Fellowship may be gained by a small group who are undoubtedly working under Jewish pressure and inspiration'.[149] Le Blount Kidd, he claimed, 'now appears is a hidden Jew' and several of the AGF's corporate sponsors 'are Jewish interests'.[150] More significantly, he leaked Pitt-Rivers copies of internal AGF correspondence and told him that his failed effort at membership had been sabotaged. The most explosive evidence was a signed statement by Wright about the proceedings of the AGF council meeting in which Pitt-Rivers' membership had been denied for a second time. Ribbentrop himself had asked for the council to reconsider the decision not to admit him, but when his name was brought forward at this meeting Mount Temple had allegedly remarked that he considered Pitt-Rivers to be 'mad', and his grandfather, the general, to have been 'notoriously mad'. Another council member then remarked that the Athenaeum Club 'deeply regretted ever admitting Captain Pitt-Rivers to membership' while a third outlandishly alleged that Pitt-Rivers 'had once shot a man in a hotel in Johannesburg and that he was "a very dangerous man"'.[151]

These were clearly slanders, and Wright and Pitt-Rivers separately considered legal action against the council members involved for their statements and for terminating Wright's employment.[152] Wright couched his objections to the organization in anti-Semitic innuendo:

City [of London] men have a perfect right to form an organization to bolster up the influence of [German economics minister] Dr. [Hjalmar] Schacht or to protect their own monetary interests, but they have no right to obtain subscriptions from the public for such an organization on different grounds...I am well aware that I have little to gain personally beyond the enmity and ill-will of a group of persons closely allied to High Finance and the great power of International Jewry, although some of these very persons are, to my knowledge, secretly hostile to the Jews from whom they accept favours – in fact, they do not seem to play straight with anybody.[153]

The AGF was rapidly splitting between its pro-business and pro-Nazi factions, and the division would only become greater as the decade progressed.

In the midst of this dispute, Britain itself was entering significant political turmoil. In January 1936 King George V had died, leaving the throne to his son, who became Edward VIII the following month. Edward's ongoing romantic entanglement with American divorcee Wallis Simpson soon prompted a potential constitutional crisis as Conservative Prime Minister

Stanley Baldwin made it clear that neither his government nor those of the Dominions would support Edward's marriage to Simpson due to his status as the head of the Church of England. This drama began to play out in the press in late 1936, and public opinion split between those sympathetic to the king and those who backed the government in its opposition to a marriage with Simpson.

In early December, Winston Churchill attended an event at the Albert Hall and promised beforehand to deliver an address related to the escalating crisis. Though he did not actually do so, the attending crowd's rousing rendition of *God Save the King* moved him to release the text of his statement to the press. In it, he explosively accused Baldwin's government of conspiring with the Labour opposition to issue the king an ultimatum. The conclusion that could be drawn from such an argument – that a 'King's Party' should emerge to back the monarch against these parliamentary machinations – raised the prospect of a serious constitutional crisis that could extend to the heart of the British political system.[154] Press opinion split, with the Beaverbrook papers backing the king while others backed Baldwin.[155] Despite a brief resurgence of support for the king among Conservative MPs in the midst of the crisis, on 7 December the House of Commons turned on Churchill and shouted him down after he posed a question to Baldwin.[156] Increasingly isolated and facing a serious political crisis, Edward abdicated on 10 December. His younger brother, the Duke of York, would be crowned George VI.

In line with his view of aristocracy and distrust of democracy, Pitt-Rivers believed that 'the King is the King' and came down firmly on the side of Edward. Following the abdication, on Christmas Day 1936 he formally resigned the commission in the Regular Army Reserve of Officers that he had held since resigning from active duty in 1919. His inflammatory letter to the War Office denounced 'a Parliamentary despotism, now styled as His Majesty's Government' and argued that 'a close and anxious study and a knowledge of past and recent changes in the sphere of domestic and foreign affairs has revealed the disquieting fact that this country is liable to be driven into civil or international war at the dictates of international powers hostile to national interests and incompatible with national survival [the Jews]'.[157] Asked to further elucidate his decision by the War Office, he was open in his derision for the British government:

> I received my first Commission from King Edward VII, my second commission from King George V, and I last held a Commission under his former Majesty, King Edward VIII, until Prime Minister, Stanley Baldwin, secured his abdication. I owe no allegiance to Mr. Stanley Baldwin, who has stated that his lips were sealed and, I perceive, that his eyes are sealed also.
>
> I may add that I was still a British officer when I was arrested by the Czech Police … without any charge being preferred against me …. When I was in Spain last November and December I was free to travel only in

the Nationalist Spain under General Franco where a British passport is, thanks also to our Government, a document of suspicion, not of security; while in Red Bolshevist Spain, so glamorously assisted and recognized by the Judaized and internationalized shopkeepers styled His Majesty's Government, my British status was treated with contempt.

The British Israelitish Raj is mud.[158]

Rejected from the AGF for his extremist views, Pitt-Rivers had now made clear his feelings towards his own government. Unsurprisingly, it was not only the War Office that would read this letter but also the MI5 agents that were now tracking his activities with diligence.[159] As Anglo-German relations continued to erode as the decade progressed, Pitt-Rivers would only increase his efforts to bring the countries together both politically and diplomatically, attracting attention from both his German connections and the British Security Service now monitoring his every move.

CHAPTER SIX

'My dear cousin Clementine'

By mid-1937, the nature of Pitt-Rivers' relationship with Becky Sharpe had become open knowledge in his social circles. While he was careful to always refer to her as his 'secretary' in correspondence, there was little secret that she was more than his assistant.[1] In addition to their travels together in Europe and their public appearances at Anglo-German Fellowship events and other venues, the pair now shared a flat in a plush Chelsea neighbourhood where they 'lived together as man and wife'.[2] One of his friends in the art world, the renowned painter Wyndham Lewis, made matching sketches of the pair and begged to use Sharpe as the model for a piece he planned to display at an upcoming show. 'She is exceedingly picturesque, and when it is done I can sell it, I know', Lewis told him.[3] Pitt-Rivers and his actual wife, Rosalind, soon entered legal proceedings against one another to end their marriage. Unlike in his first marriage, however, Pitt-Rivers fought the case in an effort to protect himself against a costly financial settlement. The resulting legal struggle stretched on for months, with Pitt-Rivers ending up paying his wife's legal costs and her being granted a separation.[4] She would soon return to her scientific work and never marry again.

Still excluded by the AGF leadership, Pitt-Rivers nevertheless continued his personal efforts to 'build Anglo-German understanding'. With Sharpe as a member of the organization, both continued to cause trouble for Conwell-Evans and Mount Temple by penning a mass of correspondence accusing the Fellowship of not representing its membership and improperly excluding Pitt-Rivers.[5] Both wrote articles for the *Anglo-German Review* discussing their experiences in Spain and Czechoslovakia, and they hoped to obtain visas to visit Spain a second time.[6] With the borders to the country officially closed, this required permission from the Foreign Office and Pitt-Rivers hoped to secure permission for a visit by arguing that he was travelling as a 'Special Correspondent' for the Catholic newspaper *The Universe* and was additionally 'proceeding to Spain on a recognized humanitarian mission in connection with medical aid and the sending of an ambulance unit'.[7] The Foreign Office was unimpressed by this implausible justification and denied the application. Questions were eventually posed in parliament to Viscount Cranborne, under-secretary of state for foreign affairs, about the decision, to which the Viscount replied simply that 'I am satisfied that there were not sufficient grounds in this case for making an exception to the regulations at present in force in regard to the issue of endorsements for Spain.'[8]

Thus, denied the chance to return to Spain, Pitt-Rivers and Sharpe made plans to spend much of the year in Germany. Among their first stops was the town of Göttingen in Lower Saxony, home to one of the country's most renowned universities. The institution had been founded by King George II in 1737 in his role as the Elector of Hanover and, as part of its bicentenary celebrations, the university authorities sought to hold an event that included as many British academics as possible. There was an obvious propaganda element to these plans: Göttingen had expelled its Jewish faculty members in 1933, including prominent physicists Max Born and Edward Teller, and attending the celebrations could be construed as an endorsement of the anti-Semitic purge.[9]

Under these circumstances, both Oxford and Cambridge declined to send official delegates and as a result the British attendees were considered to be 'unofficial' emissaries from their respective universities. Pitt-Rivers was one of several 'representing' Oxford, along with Professor Raymond Beazley, an Oxford graduate and history professor emeritus at Birmingham who was also a member of the AGF. Privately, the Göttingen organizers had tempted Pitt-Rivers with the ultimate academic recognition: the awarding of an honorary doctorate at the event. Though flattered by the offer he declined the degree, claiming later that he did not want to accept a foreign degree when Oxford had not given him an equivalent 'token of appreciation' for his contributions to the Pitt-Rivers Museum and his academic work.[10]

Following the formal celebrations and a series of academic papers given by the foreign delegates, the British delegation presented the rector of the university with a statement expressing their support for the university and hope for improving Anglo-German relations:

We the undersigned members of English and British Empire Universities, amongst the most ancient as well as amongst some of the more recent of our seats of learning, desire to convey our high personal esteem and respect, Herr Rector, for the world-renowned University of Göttingen and the tradition of high scholarship, of German culture and of science which links us in a common European heritage in pursuit of truth, learning and all we most value in civilization.

We are especially mindful of the links which bind us on the occasion of the 200th year celebration of your distinguished foundation in the history and the blood of Hanover and the history and the blood of England.

These links we would assure you are stronger and more permanent than any ephemeral clouds of misunderstanding or false appearances because they are rooted in the hearts and minds (*geiste*) of individual Germans and Englishmen. These sentiments alone, we are convinced, will bear fruit and survive, and we thank you for the opportunity and the privilege you have allowed us for giving expression to them.[11]

Pitt-Rivers was among the signatories to the statement, which was presented by Beazley in a public address.[12] The appearance of the English delegation, though unofficial and small, at the event was indeed a propaganda coup. The official Nazi Party organ in the town, *Göttinger Nachrichten*, ran an account of the English delegation's statement to the rector as its lead story, with a photograph of the participants below.[13]

The following month, Pitt-Rivers and Sharpe arrived in Paris for the General Assembly of the IUSIPP. The body was still reeling from the fiasco of the 1935 Berlin conference, and realizing that they would likely face a more hostile environment this time, the German government sent a large delegation of scientists to combat opponents of their racial views. The fireworks that would take place in the conference panels were minor in comparison to what went on behind the scenes in the Executive Committee, however. As the body's general secretary, Pitt-Rivers was required to produce a report that would be presented at the General Assembly updating the delegates on its financial status and other matters. Rather than a normal report, however, Pitt-Rivers had already circulated a draft of the inflammatory document he planned to offer. Much like his earlier memoranda to the Eugenics Society's council, it clearly had little intent beyond antagonizing and embarrassing as many of his colleagues as possible. The Executive Committee was thus left with the unenviable task of deciding whether this document should be distributed to the General Assembly at all or whether it should be suppressed as unsuitable for public consumption.

The report opened by accusing the organization's president, Sir Charles Close, of mismanaging the body's funds. Pitt-Rivers, in his capacity as treasurer, wrote that the organization's finances were 'not flourishing', while Close retorted in the Executive Committee meeting that there was nothing amiss and the finances were fine. It is difficult to know who was actually correct. The body's balance sheets were not in the negative but it was also not rich, and it is possible that Close was less concerned about the bottom line because he knew that William Beveridge at the London School of Economics would bail out the organization if needed. Finances aside, Pitt-Rivers' report went off the deep end as it went on. Given the political upheaval in Spain and the fact that there were two governments competing for power there, he argued that there was no way the country's representatives should be allowed to vote in the General Assembly. Further, there was no properly constituted Spanish National Committee to pay dues. Close directly rejected this idea and the two men openly clashed in the meeting, with Close accusing Pitt-Rivers of having 'done hardly any work at all' as general secretary.[14]

Unable to agree on the Spanish issue, the committee moved on to the other sections of the report. The Italian national committee, it emerged, had also not paid its fees. Close proposed allowing its representatives to be elected to offices like other delegates, but Pitt-Rivers objected strenuously to this 'unconstitutional' move. The discussion then moved to Czechoslovakia, which Pitt-Rivers argued should not be allowed to affiliate with the organization because German 'men of science' had been excluded from the Czech-dominated national committee. In addition, he claimed, the fact that he had been arrested in the country the previous year showed that Prague was trying to systematically oppress the truth about the situation there. He now called for the IUSIPP to send a scientific delegation to examine the Czech situation. Close and the Executive Committee disagreed on this 'purely political question' and rejected the outlandish idea.

Finally, Pitt-Rivers wrote that the 1935 Berlin conference had been a success, and 'practically all the speakers refrained from transgressions into dangerous, unscientific fields. None of the Germans touched on the Jewish question; ostensibly very strict orders had been given to them in this respect.'[15] As already noted, this was a complete whitewash of what had actually taken place and most of the Executive Committee seem to have simply ignored the statement. To conclude the meeting, a new president was elected and a new general secretary appointed to replace Pitt-Rivers, whose term had mercifully now come to an end. He was the only Executive Committee member to vote against the appointment of his own successor on the grounds that the matter had not been properly voted on.

In the end, these theatricals accomplished little more than making Pitt-Rivers look foolish. Following its negative reception in the Executive Committee, his scathing report was not officially presented to the General Assembly. Instead, he printed several hundred copies and had them placed on a table for the delegates to pick up, only to discover that they had mysteriously disappeared by the start of the meeting. Even the German delegation, including Ernst Rüdin, had abandoned him in the Executive Committee and not joined his objections to Close's actions.[16] Now ousted as general secretary, Pitt-Rivers had marginalized himself from the IUSIPP leadership much as he had already done in the Eugenics Society. He would later blame the Jews for the strife that he himself had created in the organization.[17]

From the Paris defeat, Pitt-Rivers and Sharpe returned to Berlin before heading back to Frankfurt, where Pitt-Rivers met with the mayor and other dignitaries to discuss Anglo-German relations.[18] Taking a break from their gruelling travel schedule, the pair took an August holiday in the resort town of Schlangenbad, near Wiesbaden, before returning to Frankfurt, where Sharpe bought a piano accordion and taught herself to play after hours.[19] It was not all relaxation, however, and Pitt-Rivers had now managed to secure Sharpe a place at a camp run by the local *Bund Deutscher Mädel* (BDM – League of German Girls) of the Hitler Youth. Sharpe spent nearly two weeks in the camp, later recounting her time there in detail:

> We rose at 5 a.m. – rolling out of our bunks to the sound of bells and whistles – rushed to the cold showers … – pulled on a rough uniform – smoothed our hair and formed up to marching orders in the square outside – just under the flagstaff carrying the German flag. The camp members were told off each day to raise the flag – the swastika was emblazoned across – while the rest sang the National Anthem and swore undying allegiance to Germany and the Führer.
>
> After a breakfast of bread, jam and coffee we cleaned our quarters. All morning we worked in the fields or in the farmhouses, helping the women with children, caring for those who were sick or having babies …. In the evening we had an hour before supper in which we did what we liked. Some wrote letters home, washed clothes or read books – others sat dully with their hands folded listening to their friends who stood around chattering. I usually joined this group and was the subject of much questioning – naturally enough. It always began with: 'Du bist ein Englanderin, nicht wahr?' [Isn't it true that you're English?] – It was an open sesame to the world for them, since I had been in France and South Africa as well. They were a bright and friendly lot. Politics were taboo and I too apprehensive to ask many questions in case Die Haus Muter [*sic*] were to pounce upon me. For the most part she left us alone – a formidable looking woman in a spotless brown uniform of some sort.
>
> In the evenings there was usually gruel or soup, milk vegetables and brown bread for supper – followed by a lecture on the National Socialist *weltanschauung* [world view] given by some hefty and usually rather self-important looking Gauleiter [regional leader]. After an hour's community singing, much chatter and some good-natured laughter – at the blast of a whistle we went smartly to bed, or to bunks, should I say.[20]

Sharpe's time in the BDM camp was a testimony to the status that Pitt-Rivers now enjoyed in Germany. Sharpe herself noted that her experience there was atypical and had only been arranged through Pitt-Rivers' 'determination and the esteem in which he was held in official German circles'.[21] She would later recount the experience in an adulatory *Anglo-German Review* article.[22]

There was more feting by the Nazi government ahead, and in early September, Pitt-Rivers and Sharpe received an invitation to the upcoming 1937 Nuremberg Rally. It had long been German policy to entice as many prominent Britons as possible to the annual event, in large part because Ribbentrop believed it would both impress the visitors and provide a venue for building new contacts.[23] The AGF sent an official delegation to the event and numerous members were in attendance, including Barry Domvile. With Pitt-Rivers still not admitted to the organization, his and Sharpe's invitation was arranged separately through his contacts in Frankfurt and noted that both were 'politically trustworthy' (*politisch unverdächtig*).[24]

As foreign visitors to the rally, however, they were grouped with the visiting AGF members in the British delegation, which also included the author Francis Yeats-Brown, author of *The Lives of a Bengal Lancer*. In 1935, the book had been turned into a popular film starring Gary Cooper that Hitler frequently referred to as one of his all-time favourites. He had evidently been impressed by its theme of a small number of Europeans defeating and dominating millions of non-Europeans.[25] Sharpe and Yeats-Brown were themselves 'old friends', and he had wanted her to work as a research assistant on his next book. She had declined the offer to take the position with Pitt-Rivers, fearing that Yeats-Brown was 'too temperamental'. She later conceded that Pitt-Rivers' own temperament 'wasn't a whit better – on balance I think it was even worse. I would have caught it either way'.[26]

Domvile served as the head of the British delegation, which met Hitler in a receiving line on 10 September. Protocol called for the *Führer* to greet Domvile first, but he was evidently only interested in Yeats-Brown.[27] A fellow participant, Ward Price, described the scene:

> Hitler would walk round the whole circle of his foreign visitors shaking hands with each of them in turn. I happened to be standing next to Y.B. [Yeats-Brown] and I had heard that 'Bengal Lancer' was the Führer's favourite film, so that when Hitler's A.D.C. introduced 'Major Yeats-Brown' I interjected '*Verfasser von "Bengal Lancer."*' ['The author of "Bengal Lancer."'] Hitler's face, which had hitherto worn a stiff and formal expression, became suddenly transformed. '*Was!*' ['What!'] 'Bengal Lancer!' He exclaimed. 'I have seen that film five times. It is a splendid story. Has it been translated into German?'... It was evident that he [Yeats-Brown] was the only Englishman present for whose achievements Hitler felt the slightest interest.[28]

Pitt-Rivers was also in the receiving line and was apparently greeted by Hitler briefly, though his own achievements were evidently less interesting to the *Führer*. Sharpe herself was briefly introduced to Hitler by her 'handsome blonde cousin' Unity Mitford, a frequent female companion of Hitler and an admirer of the Third Reich. Sharpe later reported that she had been unimpressed with the *Führer* during their short meeting.[29]

Thus, having been welcomed by Hitler himself, Sharpe and Pitt-Rivers attended day after day of the Nuremberg festivities, including parades, speeches by various government officials and demonstrations of German military prowess. Still the avid photographer, Pitt-Rivers took dozens

FIGURE 6.1 *Always the amateur photographer, Pitt-Rivers recorded his experiences at the 1937 Nuremberg Rally on film. He was impressed by the spectacle and got the chance to briefly meet Adolf Hitler at a reception.*

of photos of the festivities, providing a remarkable and candid view of the events, including a number showing Hitler reviewing marching SS and *Wehrmacht* soldiers.[30] The rally's official theme was labour and the celebration of rising employment since Hitler's ascension in 1933, and it is telling that Pitt-Rivers took a number of photographs of the German work brigades, the *Deutsche Arbeitsfront*, that marched at the rally carrying shovels rather than rifles. He no doubt would have compared this spectacle to his own views of the need for British agricultural reform and renewal. The rally also provided an opportunity to meet a variety of Nazi officials first-hand, and Pitt-Rivers attended a reception held by SS chief Heinrich Himmler among the other events.[31]

While Pitt-Rivers and Sharpe appear to have attended most of the rally's events, not all members of the British delegation treated the visit with the same reverence. During a review parade of the *Wehrmacht*, one of the AGF's official delegates opted to view the parade from the window of a beer hall, 'through the end of a beer tankard' and eventually became so intoxicated that he had to be carried to the delegation's bus 'on the shoulders of a young S.S. man in blissful oblivion'.[32] The same delegate had been 'almost every day... very much the worse for drink' and on this occasion ended up 'sprawled all over the seat, finally subsiding on the floor' before reaching the delegates' hotel.[33] While no doubt an entertaining spectacle to some, for Pitt-Rivers and Sharpe this was 'humiliating conduct' that hardened both even further against the AGF.

The AGF itself was by now undergoing a formal division that was playing out, in part, at the Nuremberg Rally. Disappointed with the AGF's measured tone towards Nazism and its emphasis on recruiting elite members, Domvile had founded a new body called The Link that he was touting within the British delegation. While the AGF's stated goal was the building of Anglo-German relations generically, The Link would include members who were openly and vocally pro-Nazi. The AGF recruited primarily elites – businessmen, bankers, members of parliament and aristocrats – but The Link would be open to members of all backgrounds and, indeed, most members were ordinary people from areas outside of London. The organization's main objective was building Anglo-German relations by arranging trips to Germany and other forms of cultural exchange. In essence, it was a down-market version of the AGF that remained unconstrained by the strictures of establishment 'respectability'.[34] The AGF and The Link themselves had a strange institutional relationship: while both had an interest in remaining separate from one another, The Link became a frequent topic of discussion in *The Anglo-German Review* and received a great deal of coverage in the periodical. This was no doubt because its editor, C.E. Carroll, was also a founding member of The Link and was becoming increasingly open about his own anti-Semitic views in the pages of the *Review*.[35] The AGF was losing control of even its own periodical.

Pitt-Rivers was a natural recruit for The Link. His views on Hitler and National Socialism were now well known, and he had spent extensive time in Germany with friends who would make useful contacts for the fledgeling organization. However, he proved difficult to convince to join, in part because of a serious misstep on the part of its leaders. Professor Raymond Beazley, with whom Pitt-Rivers had attended the Göttingen celebrations earlier in the year, wrote to him in August sharing the early plans for The Link and asking for his views, assuming that Domvile had already consulted him about the idea and invited him to join. This was, of course, the first Pitt-Rivers had heard about it, and Beazley had inadvertently insulted him by asking his views of the scheme before Domvile had even mentioned it to him.[36] Infuriated, Pitt-Rivers evidently spent much of his time in Nuremberg vocally criticizing The Link and bombastically claiming that he had come up with the exact same idea years before.[37] He told Beazley that The Link's origins in the AGF might doom its prospects from the beginning, citing its very creation as part of the 'Jewish plot' he now evoked to explain virtually everything he disliked or with which he disagreed:

> It appears to me that the old intrigues of the 'Palestine Potash' group in the A.G. Fellowship have been continued in order to ensure that the Zionist forces shall remain in virtual control. It is announced that Lord Mount Temple has 'approved' of the new extension [The Link] and is duly advertized with Conwell-Evans in recent numbers of Carroll's

[Anglo-German] Review. Conwell-Evans stayed with the Ambassador von R[ibbentrop] in Germany before the Partaitag [Nuremberg Rally]. I was greeted with marked frigidity by the Ambassador at [SS chief Heinrich] Himmler's reception. This would be of no particular significance where [*sic*] it is not for the fact that having been pressed by several members of the Fellowship to join it I was refused membership, when eventually I consented to join, and firmly excluded when the Ambassador himself intervened to seek my admission. I was afterward asked to lunch by Mount Temple to discuss matters and when I wrote to him after that lunch to ask him whether the remark he was alleged to have made to the solicitor of the Fellowship – according to this person himself – (the remark was a criminal libel) was in fact made by him, I received no reply .... With regard to any sort of collaboration with the A.G.F., as at present controlled, that is, as I feel sure you will agree, obviously impossible. The true object of M.T., Tennant and Conwell-Evans is to corrupt or crush National Socialism in Germany, just as it is the policy of our own Foreign Office.[38]

This was obvious nonsense on several levels. Pitt-Rivers had been rejected from the AGF not because of some grand conspiracy but because his views were too extreme for a body struggling to maintain its respectability despite being an obvious tool of Ribbentrop and the German Embassy.[39] Beazley assured Pitt-Rivers that 'we will get you into the A.G.F. a little later' and urged patience, but at this stage Pitt-Rivers was too far beyond the pale for this to be a realistic hope. Sharpe published the account of her time in the BDM camp in the December 1937 edition of the *Anglo-German Review*, but it would be the last time either she or Pitt-Rivers provided material for the periodical.[40]

Pitt-Rivers now embarked on a new publication project of his own that he believed would make a major contribution to Anglo-German relations. He had already abandoned the prospect of producing a new edition of *The Clash of Culture*; any such undertaking might require additional fieldwork, which was out of the question, and, besides, he had already re-tasked Sharpe to work on a new edition of *The World Significance*, of which he now proposed to extend the chronological scope from 1917 to 1937.[41] This new version would be oriented towards examining 'Jewish influence' on world affairs since the 1920 edition had come out.

At the same time, however, Anglo-German relations were themselves deteriorating. In March 1938, Austria was annexed into the Third Reich, with German troops crossing the border on 12 March and reaching Vienna three days later. Rather than resistance, they were met with cheering crowds. The annexation – called the *Anschluss* – was shocking to many in Britain, particularly when a less-than-free plebiscite approved the formal annexation with more than 99 per cent approval. A headline in *The Times* referred to the event as the 'rape of Austria', while the correspondence section of the paper debated whether Britain should take action or remain free of European entanglements lest they lead to the outbreak of another war.[42]

Pitt-Rivers, as was typical, took the German perspective. The *Anschluss* was not only justified, he believed, but it was also a major victory for Hitler over Bolshevism and the Jews. 'I think that Germany and Adolf Hitler are to be complimented on the Austrian coup', he wrote a friend the day after German troops arrived in Vienna. 'The country was very near to civil war a few months ago. My effort to promote Anglo-German understanding is very relevant just now.'[43] The same day, he told a German contact that the political earthquake was a positive development: 'Contrary to opinion in this country I and some of my friends look upon it as a step nearer to international peace than war', he wrote.[44] He was accusatory in his scorn towards those who did not welcome the move: 'the only section of the population who do not appear to welcome the union is the Catholic-Communist-Jewish one!' he claimed.[45]

Another gesture of support for the *Anschluss* was soon to follow. In late June, Pitt-Rivers wrote to Hitler himself, via the German Embassy in London, to express his admiration and send a copy of a recent speech in which he expressed hope for a new understanding between Britain and Germany:

> Allow me, an old British officer and sincere friend of Germany, who had the honour of being received by Your Excellency during the last Parteitag at Nurnberg, to transmit to you the accompanying special copy of my address 'Der Friede Welcher Hohe ist denn alle Unvernunft' ['The peace which surpasses all misunderstanding'], including also the address we English University graduates presented to the University of Gottingen on the occasion of its 200-Jahrfeier [Bicentenary Celebration].
>
> May I also express my sentiments of profound thankfulness that the Anschluss with Austria has been accomplished under your leadership without bloodshed and with the rejoicing of all the German and Austrian peoples.
>
> With many of my friends I recognize that the true opinions and sentiments of Englishmen sympathetic to National Socialism are seldom allowed to reach you owing to the control of all organs of public opinion in the press and parliament by those hostile, alien and international powers which Friedrich Nietzsche exposed half a century ago...
>
> With profound admiration, allow me to remain, in devotion to the cause of a New Europe.[46]

It is unlikely that Hitler ever saw the letter personally, though the German Embassy sent a perfunctory reply expressing the *Führer*'s thanks for his friendship.[47]

The letter was a telling expression of Pitt-Rivers' views as the Second World War approached: National Socialism, he now believed, promised to save Europe from the degeneration of Bolshevism and the corrupting influence of the Jews. Opponents of Hitler and National Socialism, he believed, were either themselves Jews or had been deluded by Jewish interests, as he had wildly suggested about the AGF. Jewish economic control, he now believed, was at the centre of the conspiracy, and in late 1938 he forbade his son Julian from attending Cambridge to read the economics tripos on the grounds that it was no more than 'judeo-financial magic' and 'worse than a shocking waste of time'.[48] Maintaining Anglo-German relations, in his mind, now meant defending both Germany and Britain against these influences. This was a task for which Pitt-Rivers believed himself uniquely qualified: after all, in his view *The World Significance* was 'analytic in retrospect and prophetic in prospect' and had accurately predicted events since its writing.[49]

In the aftermath of the *Anschluss*, the next European crisis was already heating up by summer in an area with which Pitt-Rivers was particularly familiar. Konrad Henlein's Sudeten German political party had now embraced violent measures against the Prague government, influenced in part by German financial support for the cause and Hitler's increasing desire to dismember the country. Long-sought concessions by Prague towards the German minority, including admission of German-speakers to the civil service, were obtained in 1937, but these were now no longer enough to sate Hitler's thirst for Czech territory. Henlein now played the part of the pawn by making increasingly unreasonable demands on Prague at Hitler's behest.[50]

This was an extremely dangerous game. In May 1938, the Chief of the Army General Staff informed Hitler that Germany could probably not expect to win a war in which the Germans would likely face not only the Czechs but also the French and British if those countries decided to join the conflict. Thus faced with internal disagreement from the military alongside international opposition to a German annexation of the Sudetenland, Joseph Goebbels unleashed the Nazi propaganda apparatus. Throughout the summer of 1938, German newspapers – with many foreign publications following suit – were filled with accounts of alleged Czech atrocities against the German minority, including stories describing the murder of women and children by the Czech police and threats of gas attacks on German-speaking settlements. The fact that Czechoslovakia was allied with the Soviet Union presented German propagandists with an easy argument that the Czechs were working as the agents of Bolshevism.[51]

Pitt-Rivers now saw himself as an agent of destiny in the unfolding Sudeten crisis. His 1936 visit to the country, his conversation with Henlein and his arrest by the Czech police were now no longer a matter for discussion merely in the *Anglo-German Review* and in confidential letters to the Foreign Office: the time had come to make the case on a larger stage. The stakes, in his mind, were now no lower than securing European peace and maintaining the struggle against communism. Abandoning his other publication projects, Pitt-Rivers set about writing an account of his experience and research findings from the 1936 trip. The resulting book, *The Czech Conspiracy: A Phase in the World-War Plot,* would achieve Pitt-Rivers a level of popular fame he had never enjoyed in his previous work.

The Czech Conspiracy was, in many ways, the intellectual culmination of Pitt-Rivers' career. Blending standard anti-Semitic tropes with his usual anti-communism, Pitt-Rivers evoked his scientific credentials to insist that the book was a work of demography and empirical research. In the first pages, he claimed that his purpose in writing was to 'set out calmly, dispassionately, and briefly the principal facts known to me, by personal observation and experiences, by twenty years' study of these problems in Central Europe, and by my recent adventures in the country which is the Tinder-Box of Europe'.[52] In accordance with the scientific legitimacy he claimed to bring, the book included detailed demographic and historical maps of the region, tables with the statistics that he and Sharpe had gathered during their trip and pages of photographs illustrating the poverty and suffering they had observed in the country. The fact that these had been taken nearly two years before and might reasonably be argued to no longer represent reality was apparently not a concern.

For all the solid demographic data that might have been present, however, Pitt-Rivers' wider agenda could be easily discerned, and it beguiled his claim to scientific objectivity. The book's tone was inflammatory from its first pages, opening with an account of Czechoslovakia's origins that accused 'The Bolshevik-internationalist destroyers of our war ally, Russia, and the Bolsheviki's powerful financial friends in the United States of America' for its creation after the First World War.[53] Czechoslovakia itself was a 'monstrous political fraud perpetrated upon the helpless minority peoples of the dismembered Austro-Hungarian Empire' and a 'vassal state' of the Soviet Union committed to luring Europe into another cataclysmic world conflict.[54] German minorities all over Europe had been the great losers of the Treaty of Versailles and were now oppressed in virtually all corners of the continent, he claimed, and only Hitler and Henlein could right these wrongs.

Pitt-Rivers committed these arguments to paper in the summer of 1938 and the book went to press on 26 September 1938.[55] By this time, the Czech crisis had already reached a breaking point. On 12 September, Hitler gave a Nuremberg Rally speech calling for the Sudeten Germans to be given independence from Prague and promising war in the event that his and Henlein's demands were denied. Violence broke out in German-speaking areas as Henlein's supporters tried to provoke police repression and precipitate military intervention from across the German border.[56] Just as it had been in the summer of 1914, Europe was once again potentially on the brink of a world war. Following Stanley Baldwin's resignation as Prime Minister the previous year, Neville Chamberlain had assumed the office and knew that Britain was in no condition to fight a large-scale war either financially or militarily.[57] At the height of the crisis, the government issued gas masks to civilians; buildings in London were fortified with sandbags; and parks were dug up with trenches.[58] Yet there was the pressing question of whether Britain should allow itself to be dragged into another European war, as it had in 1914. Chamberlain laid out this case against intervention to the British people in a radio address in late September: 'How horrible, fantastic, incredible it is that we should be digging trenches and trying on gas-masks here because of a quarrel in a far-away country between people of whom we know nothing', he told the public.[59]

Sharpe and Pitt-Rivers had a unique perspective on the crisis, having personally visited the Sudetenland and met Henlein, albeit briefly. His views would soon be public in the first edition of *The Czech Conspiracy*, but hers were more mixed. As she later recalled of the period:

> I was constantly bewildered by opinions and events. What was one to believe? There were some who said that Austria had welcomed the *Anschluss* – others that she had been raped. I knew people who spent many months in Germany, feted by Party officials – as Jo had been – who were quite free to move where they pleased. Some of them said with genuine conviction that the Nazi programmes were the finest efforts at social reconstruction the world had seen for centuries and that Hitler had made the German people united and happy at last. I knew of others, who had Jewish connections, refugees, who had suffered great distress of mind as well as physical persecution at the hands of the ubiquitous Gestapo...
>
> Many of the women w[ith] whom I spoke remained silent. If their husbands and sons saw reason for the coming conflict and might even justify it, they could not. Even those who wanted war would be left – as the women had always been throughout history – to face the thing alone at home. There were times when it seemed to me that it did not matter very much what one thought – it was coming anyhow.[60]

In late September, Chamberlain personally flew to Munich to meet a delegation of German, French and Italian representatives and heads of state in an effort to avoid war. Notably, the Czechs were not invited. Hours of negotiations culminated in the signing of an agreement on 29 September to hand over the Sudetenland to the Reich, thus dismantling Czechoslovakia. Hitler then signed a declaration stating that the British Empire and Germany would never go to war against one another. Returning home, Chamberlain waved the signed document in front of cheering crowds, declaring 'peace for our time'.[61] In March 1939, both the British and the French looked the other way as Hitler marched into Prague in contravention of the agreement.

With the signing of the Munich Agreement just three days after it had been published, the first edition of *The Czech Conspiracy* was now woefully outdated. Pitt-Rivers' arguments about the plight of the Sudeten Germans were now of historical rather than political interest; their annexation to the *Reich* had already been achieved. He immediately set to work writing additional chapters to make the book timely. The largest section he added was replete with quotations from Henlein and Hitler justifying the annexation of the Sudetenland, and a long digression denounced the subsequent arrival of refugees from Central Europe in England. His most withering criticism, however, was reserved for the British government. After the Munich Agreement had been signed, he claimed, debate in Parliament revealed 'how disappointed many prominent politicians were that there was no war', including future Prime Minister Winston Churchill.[62] Refugees from the Reich, many of whom were Jewish, were already flooding into the country, he continued, and the government was planning to admit more, which would only weaken the British Empire and potentially lead to revolution.[63] The only way out of the continuing crisis was for Britain to give Germany a free hand in European affairs:

> What is the solution, now and in the future? It is simple, and becomes every day more obvious. Remove British interference from the affairs of Central Europe and the rest follows. Let us face realities and put our own house in order. Germany will and must rescue her own German children, Poland will rescue the Poles, Hungary the Magyars. This, after all, is only that self-determination which we guaranteed.
>
> Let the Czechs be given an area of their own or return to Bolshevik Russia ...
>
> Fight we will if fight we must, not against Germany for daring and being strong enough to look after her own sons, but against the enemy in our own midst! Thus, and thus alone, can England maintain peace and her own honourable obligations.[64]

The 'enemy in our own midst' was, as usual, Jews and the communists. The book concluded with the publication of a letter to the editor of *The Times* praising the Munich Agreement as 'nothing more than a rectification of one of the more flagrant injustices of the Peace Treaty [of Versailles]. It took nothing from Czecho-Slovakia to which that country could rightly lay claim and gave nothing to Germany which could have been rightfully withheld'. Along with Pitt-Rivers, the signatories included Raymond Beazley, Barry Domvile, C.E. Carroll (editor of the *Anglo-German Review*) and Lord Redesdale (father of the Mitford sisters and Oswald Mosley's father-in-law).[65]

Actually appearing in print in early 1939, the updated version of *The Czech Conspiracy* quickly gained Pitt-Rivers the greatest international fame he had ever enjoyed outside academia. While reviews of *The World Significance* in

1920 had been mixed and its reception muted by the fact that Pitt-Rivers left for Australia just weeks after its publication, he now ensured that this new work would be as widely read as possible. Complementary advance copies were despatched to virtually every German Pitt-Rivers had ever met, including Alfred Ploetz and the new German ambassador in London, Herbert von Dirksen.[66] The French anti-Semitic author Georges Batault received a copy, as did Gino Gario, an Italian author and journalist living in London who Pitt-Rivers hoped would facilitate an Italian translation of the work.[67] Copies were sent to leading scientific journals, including *The Quarterly Review of Biology*, edited by his old friend Raymond Pearl, along with popular journals and newspapers.[68] Sending a copy to Prime Minister Neville Chamberlain, Pitt-Rivers enclosed a note claiming that his 1936 arrest in Czechoslovakia had been because 'they feared I would publish the *truth*'.[69]

The marketing campaign was a success. Letters came in from admirers of the book around the world, including Germany, the empire and the United States. A senior Royal Navy officer wrote to say that he had bought several copies to distribute to his friends and asked for help rebutting criticism in a local newspaper.[70] Pelley Publishers, an American publishing house associated with William Dudley Pelley and the anti-Semitic Silver Legion, wrote to enquire if Pitt-Rivers could help provide data on the number of Jews living in the United States, believing that 'they are coming into this country now at an unprecedented rate'.[71] Anti-Semitic Christian leader Gerald B. Winrod wrote from Kansas to ask Pitt-Rivers whether he could provide evidence that former Czechoslovakian President Edvard Beneš was Jewish and telling him that 'An awakening regarding Jewish control is taking place in America. Those of us who are leading in the fight are, of course, encountering defiant opposition. Our press and radio appears to be largely dominated by Jewry.'[72]

Sales of the book were slow but 'improving steadily', as Pitt-Rivers told his son, but its purpose was not to make money but spread his message.[73] Its success had been hindered only, he claimed, by being 'boycotted by the "Times" and the Anglo-Jewish press'.[74] With its publication, Pitt-Rivers had nailed his political colours to the mast. Though he had explicitly described himself as a scientific investigator who was 'strongly, even passionately, biased – but only in favour of the truth', he was open about the fact that he had more than 'purely academic interests' in mind.[75]

This increasing obsession with Hitler and National Socialism deeply troubled his old friend Oscar Levy. After a long silence, the two men dramatically met by chance on a ferry to Newhaven in late 1938. Though the meeting was brief, it concerned Levy greatly. Writing to his old friend days later, Levy was blunt:

> It was hardly twenty minutes, but it proved to me, that we have much to tell each other…I know you and I are going different ways. I have

remained an Orthodox Nietzschean, even *more so* – that is to say, I refuse to mix Nietzsche up with present-day politics. You have become an Aryan- Nietzschean – yet I hope, that you will still see some difference between yourself and the masters of the Third Realm. Never mind – we have once been comrades – I must and shall not, forget that…I am not afraid for Nietzsche's future: he will go on…much blood and ink will flow in the meantime. His hour has not yet come, and it will not come through Nazis but through a combination of the best in all nations.[76]

Levy's criticism left Pitt-Rivers undeterred. In a draft article intended for publication in Germany, he now implausibly claimed he had helped lay the intellectual groundwork for Hitler's rise to power:

I therefore confess that seventeen years ago, in ignorance of the will and the ideas (*Wille* and *Vorstellung*) germinating in the mind of Adolf Hitler…I was vividly aware that the existing political and economic structure of society was ceasing to function and was bankrupt…I was one of the very few who cherished a new determination and hope which is now being expressed in the aspirations of the various 'functional' national movements of integration that will ripen, as I firmly believe has happened in Germany, into a new European renaissance.[77]

Still barred from the AGF but increasingly emboldened in his criticisms, Pitt-Rivers continued to court trouble there with Sharpe's help. In June 1938, the pair attended a meeting headlined by Yeats-Brown in which he 'asked some penetrating questions of the speaker…when some one mentioned the persecution of the Jews', evidently to the embarrassment of Foreign Office official Rex Leeper.[78] T.P. Conwell-Evans subsequently wrote Sharpe a letter threatening in no uncertain terms that 'if you persist in bringing to the meetings of the Fellowship guests whose applications for membership have been rejected, you will be invited to attend a meeting of the Council, when the question of your membership will be considered'.[79] Sharpe retaliated by now trying to muster the pro-Nazi factions within the AGF against its leadership. As she told Margaret Bothamley, one of the most virulently pro-Hitler members of the organization:

In any case Margaret we have dallied about enough over the A.G.F. business and I do think we ought to make an effort to rally enough members in order to call an extraordinary general meeting. Do you think it can be done? One-tenth of the membership is required to call an Ex.[ecutive] Gen.[eral] Meeting. They have been unwise enough to throw out a challenge and we have every opportunity for a come back. We are perfectly in order demanding that the members of the Fellowship should have some voice in its administration.[80]

These machinations came to nothing beyond Beazley and Domvile stonewalling her and giving assurances that they would arrange for Pitt-Rivers to be admitted to the AGF 'before long'.[81]

In early November, Germany exploded into days of anti-Jewish rioting following the assassination of a diplomat in Paris. The violence, known as the *Kristallnacht* (Night of the Broken Glass), was directly endorsed by Hitler and inflamed by Goebbels' propaganda machine. Led by the SA (*Sturmabteilung*), mobs burned down synagogues and raided Jewish shops and homes. The official death toll was given as ninety-one but the real number was much higher. Tens of thousands of Jewish men were arrested and confined in concentration camps.[82] While the violence had in fact been carefully planned and coordinated, the German press presented it as a spontaneous uprising against 'international Jewry' and outrageously claimed that it had been restrained and 'kept within definite limits'.[83]

The pogrom was a political nightmare for the leadership of the AGF. The organization sent a letter to its membership and to *The Times* denouncing the violence as having 'set back the development of better understanding between the two Nations'.[84] Lord Mount Temple resigned as chairman 'because of the treatment of the Jews in Germany and the attitude of the Germans toward the Catholic and Lutheran communities'.[85] The pro-Nazi faction welcomed his departure, with Carroll sarcastically remarking, 'Mount Temple's exit has been a big blow for the Fellowship I imagine. What will their tea parties be without him now!'[86] Domvile himself resigned in early 1939, claiming that Conwell-Evans ('a dirty little twister') was attacking The Link and he could no longer go on. Their past conflicts now resolved, Pitt-Rivers offered to help him form a local chapter of The Link in Dorset and, using a naval analogy, told him the time was nearly ripe for an open assault on the opponents of Nazism:

> I am delighted to hear that you are going to engage the enemy fleet. As soon as you give the order 'Sink at sight', I am with you. I am not so keep about this Q-boat warfare, of disguised merchantmen [that were heavily armed to lure attackers and then destroy them]. I prefer fighting with the flag *showing*. It is surely better tactics now that the enemy is engaged.[87]

Pitt-Rivers' final salvo before the outbreak of war would be his most outrageous. As the Sudeten Crisis unfolded in 1938, the government had begun creating a civil defence plan that would be implemented in the event of war. Assuming that London would be an immediate target for German air raids or gas attacks, the plan contained clauses providing for the evacuation of children from the city and their billeting in private homes in the countryside.[88] In the event that these plans were actually implemented they would involve the movement of more than one million children as quickly as possible, and as a result the government spent much of 1938 and

1939 convincing parents to register their children, issuing necessary identity cards and actually practicing the evacuation.[89]

This plan was also naturally dependent on the willingness of people in rural areas to house evacuees. Initial surveys reported that rural residents would be willing to voluntarily house at least 250,000 visitors, but cooperation began to wane when people realized that war might actually happen and they might soon have long-term houseguests.[90] As Carlton Jackson has observed, in most cases the further a village was from London, the more accepting it was of evacuees.[91] The government eventually made it compulsory for rural householders to accept evacuees, and while most owners of large estates and manor houses did so willingly, others needed more convincing.[92] The fact that many of the evacuees would presumably be arriving from the slums did little to convince the holdout rural gentry and villagers that they should be welcoming.[93]

Pitt-Rivers was among the critics, arguing that civilian billeting was illegal, as was the canvassing of rural areas to determine their capacity for housing evacuees.[94] Mass evacuations would cause a health crisis and outbreaks of venereal disease, typhus and diphtheria, he claimed.[95] Further, he wrote, many of the evacuees would not be British at all:

> We are strongly supported in the contention that the scheme to billet unaccompanied children is a cloak and a pretext to break up family life on behalf of Communist, Jewish, and other undesirable elements among the so-called 'refugees' from abroad and from London, admitted on 'travel documents'....The great bulk of responsible English parents of all classes either in towns or in the countryside would never voluntarily consent to be parted from their children in order to send them to strangers, any more than decent families of all classes will consent to receive, indiscriminately, alien or strange children into their own homes and mix them with their own children.[96]

The idea of billeting itself, he concluded, was part of a communist plot:

> It has long been part of the Bolshevik plan to break down the unity of the family and to separate children from the control and custody of their parents. Hosts of 'refugee' Communists hope to do this in this country as they have in Soviet Russia. This is why they support the scheme to billet 'refugee' children on kind-hearted strangers in the countryside, who do not know what they are up to.... The efficiency of every home would be destroyed by inflicting a number of strange children upon every householder.[97]

These sentiments were published in a short pamphlet polemically titled 'Your Home is Threatened!' in 1939 by the Dorset No Refugee Billeting

Campaign of which he was the head (and, possibly, the only active member).[98] Even before its appearance in print, Pitt-Rivers had attracted press attention with his vocal condemnations of billeting. In a statement quoted widely in the press, he argued that even in the event of war there was no danger to the civilian population:

> Speaking as a soldier who knows something of modern warfare, apart from experience as a staff captain in the last war in charge of billeting in England, Ireland and France, the whole billeting scheme is ridiculous. Every expert knows that a congested civilian population does not present a military target, and that aerial warfare plays a very small part.[99]

This claim, along with his denunciations of billeting children, was met with widespread mockery. The tabloid *Daily Sketch* quoted his musings, acerbically commenting 'We are not so expert as Captain George Henry Lane Fox Pitt-Rivers, but we feel that a congested civilian population is still some kind of a target.'[100] The *Evening Standard* ran a Colonial Blimp cartoon with a character declaring, 'Gad, sir, Pitt-Drivels is right. This billeting of children idea is damned nonsense. The poorer classes must be lacking in decent family instinct not to want their brats blown up with them.'[101] Ignoring this derision, Pitt-Rivers tried to take direct action against the billeting plans by drawing up a petition opposing the scheme and sending it around his estate to be signed by his tenets, who, according to MI5, were told that the refugees would be 'East End Jews, Polish Jews and Czech Communists'.[102]

Perhaps sensing that none of this was heading in a good direction, Becky Sharpe decided to make her exit from Hinton St Mary in mid-1939. She had worked for Pitt-Rivers for nearly five years, during most of which she had been his companion rather than merely a secretary. In December 1937, MI5 noted that she was living in the Hinton St Mary house and 'is hoping to become his wife in due course'.[103] She had since changed her mind, as she later recalled:

> By the middle of 1939 I was getting restless and decided that I must go. The pattern at Hinton had become too repetitive – as all patterns do. I have learnt a great deal from Joe [*sic*] – from his work and the many contexts in which he functioned – but it was time to move on – 'to another dimension'. I was too afraid to let him know of my decision, for he was clever, crafty and ruthless – and I knew he would do everything to force me to stay. The truth was, he had become dependent upon me and I knew it. I was too immature to realize the extent of it then but, since I felt neither morally nor legally bound to remain, nor under any obligation to him, why shouldn't I go?[104]

FIGURE 6.2 *Happier times for 'Becky': Following her departure from England and the end of her relationship with Pitt-Rivers, Catherine D. Sharpe met Lance Taylor on board the ship to South Africa. They married after arriving in Cape Town just days after the outbreak of the Second World War.*

Sneaking her bags out of the house piecemeal over weeks and stashing them with a friend in London, she secretly booked passage on a ship to South Africa and left Pitt-Rivers a farewell note without a forwarding address.

Before leaving, Sharpe confided to her mother that she feared Pitt-Rivers would book a berth for himself and his son Michael on the same boat if he discovered her plans, resulting in 'a horrid scene and scandal on her arrival' and leading to the secrecy.[105] She had also cleverly managed to secure a financial settlement from Pitt-Rivers for the several years of wages he had not paid her after their romantic entanglement had begun.[106] Discovering her deception, Pitt-Rivers blustered about sending Michael to South Africa to confront her, but in the end evidently did nothing.[107] He probably never saw her again. Becky Sharpe – now resuming her real name of Catherine – met a doctor on the ship named Lance Taylor and was enthralled with him. They married at the Cape Town Magistrate's Court after disembarking in September 1939, just days after the start of the Second World War.[108] Catherine Taylor, as she was now called, would soon surpass the fame and accomplishments of her former employer.

With his former assistant now out of the picture, Pitt-Rivers needed new aides to help with his workload. Before Sharpe's departure he had hired another secretary named John Coast to help with the management of his estate and the anti-billeting campaign. Coast was a mysterious young man who had been recommended to Pitt-Rivers by the author Henry Williamson, now famous for the novel *Tarka the Otter* but known for his fascist and Mosleyite leanings at the time.[109] Coast was with Pitt-Rivers only a few months but 'did a host of mischief' while living in the Hinton St Mary house. His political views were evidently so extreme Pitt-Rivers himself described him as 'a very keen pro-Fascist, Nazi fanatic' who associated with a 'mob' subscribing to 'the more vehement and less responsible type of anti-Semitism'.[110]

Coast had been responsible for driving around the estate threatening tenets who did not sign the anti-billeting petition (which Pitt-Rivers later denied ordering him to do, though he did not deny writing the petition) and ill-advisedly harangued the Chief Constable of Dorset with 'violently pro-Nazi, anti-British, anti-Jewish and anti-Roman Catholic' rhetoric that was 'very near to being treasonable'.[111] His activities were decidedly suspicious and Pitt-Rivers claimed to have caught him rifling through his personal papers and documents related to his marital difficulties. He also appears to have carried on a brief romantic liaison with Becky Sharpe. Pitt-Rivers believed him to be an MI5 *agent provocateur* or informant, but there is no evidence in the available files to support this claim. Whatever his motivations, Pitt-Rivers woefully reported that he was 'in my house doing me a lot of damage' and was an 'embarrassment'.[112] At some point Coast was either fired or left his employment, though not before Pitt-Rivers had offered to introduce him to Rolf Gardiner, suggesting that he had been pleased with his activities at one point.[113]

Pitt-Rivers then hired Gladys Fortune, a young London secretary, as his assistant. Fortune would become nearly as embarrassing as Coast, though somewhat less destructive. She had been recommended by one of Pitt-Rivers' solicitors, George Richard Tildesley, who the Security Service suspected of harbouring fascist sympathies himself.[114] Fortune had been engaged to a German *Luftwaffe* officer who fought in the Spanish Civil War and she subsequently became involved in various far-right organizations. Pitt-Rivers claimed she had once been assaulted by 'Jew Communists' and had a 'tendency to hysteria' as a result of her past, along with a tendency to make inflammatory and sometimes violent remarks.[115]

By the outbreak of war, Pitt-Rivers had thus surrounded himself almost exclusively with fascists and Nazi sympathizers. Fortune was living in his house; Tildesley was running many of his legal affairs; Barry Domvile was a regular visitor; and he was hosting parties that included Major-General J.F.C. 'Boney' Fuller, a brilliant military theorist and occultist who had made significant contributions to armoured warfare tactics in the 1920s before developing affinities for the BUF and Hitler. In May 1939, Fuller attended Hitler's 50th birthday party and was perhaps the most openly political British general of the interwar period.[116] Like Pitt-Rivers and Domvile, he professed a deep mistrust of politicians and general scepticism towards democracy.[117] Among Pitt-Rivers' correspondents were H.W. Luttman-Johnson, the founder of the January Club, and Lord Alfred Douglas, the former lover of Oscar Wilde who had turned to anti-Semitism late in life. Pitt-Rivers sent him copies of his various writings on Nazism, and Douglas replied that he agreed with '*all*' of Pitt-Rivers' views and also found Mosley's ideas to be 'very sound'.[118]

Despite the signing of the Munich Agreement that was supposed to secure peace, international events were now careening quickly towards war. In May, Hitler announced to his inner circle and military commanders that an attack on Poland was necessary to secure the German-speaking city of Danzig and the future of the Reich. If Britain and France were to intervene to defend Poland, he went on, a full-scale European war might result.[119] At the same time, Hitler was quietly courting Soviet leader Joseph Stalin under the realization that avoiding the outbreak of a catastrophic multi-front war after the invasion of Poland meant keeping the USSR out of the conflict. Stalin, for his part, knew that his country would have no chance against the German army in 1939 and needed to play for time. On 23 August, Ribbentrop signed a formal non-aggression pact with his Soviet counterpart, Vyacheslav Molotov, in which the countries agreed to peace with one another for ten years. Secret clauses provided for the future division of Eastern Europe and Poland between Soviet and German control. Neither Hitler nor Stalin believed the treaty would actually survive for the full ten years, but it was temporarily in both their interests to sign it.[120]

This non-aggression pact between the virulently anti-communist Nazis and the world's first lasting communist state was predictably shocking to many. No accounts of Pitt-Rivers' views of it survive, but there is no evidence to suggest he was particularly outraged by it. He probably blamed his own government for driving Hitler into the arms of the hated Bolsheviks through its refusal to deal with Germany directly. He later wrote that the British government had no strategy at all beyond antagonizing Germany:

> It had not been decided upon [in 1936] whether we were in honour bound to defend 'gallant little Finland' against the Tartar hordes of 'totalitarian' Russia, or to denounce the 'aggression' of the Finnish War Party against the democratic security of the Soviets, or to assist the Russians to liberate the Poles by incorporating Poland into Russia together with the other Baltic republics, or to assist the Poles to liberate Danzig and East Prussia by incorporating those territories into Poland, or to assist Czechs and the Yugo-Slavs against the territorial expansionist aims of Greater Germany, or to fortify France against the aggressive designs of Nationalist Spain.... Eventually, however, by a brilliant diplomacy in our conduct of foreign affairs, and in pursuance of our faded policy of 'balance of power', we

finally achieved a unity of national purpose and confounded all prophets by doing all these things in turn....[121]

With the Soviet Union no longer posing a threat to his plans – at least for now – Hitler saw the opportunity to move into Poland. From the middle of the year, the Nazi press had been spreading tales of atrocities against the ethnically German population of the country, much as it had done in the period before the Sudeten Crisis.[122] As war approached, the British and French governments reasserted their commitments to defend Poland in the event of German invasion. Hitler believed the British would not be willing to go to war over the country, and early in the morning of 1 September, the SS launched a false flag assault against a German radio station in Upper Silesia, leaving behind the bodies of concentration camp inmates dressed in Polish uniforms to suggest that the Poles had been responsible for the attack. Later in the day, Hitler addressed the Reichstag to announce that the Polish military had violated the German border and demanded a military response. The full-scale invasion of the country began soon after, and with it, the Second World War.[123] On 3 September, Britain and France issued an ultimatum to Hitler demanding German withdrawal from Poland. When this was ignored, both countries declared war. Addressing Nazi Party members, Hitler blamed 'our Jewish democratic global enemy' for misleading Britain into the conflict.[124]

Pitt-Rivers' grand notions that his personal efforts could help prevent the outbreak of war had now been revealed to be dangerous delusions at best. Britain was once again at war: parks were again dug up with trenches; sandbags lined the exteriors of government buildings; three million children were evacuated from cities and moved into the countryside; and soldiers were billeted in private homes around the country.[125] Some of his old friends on the far-right now disavowed their past views and committed themselves to the war effort. Sir Arnold Wilson MP, with whom Pitt-Rivers had corresponded about Franco and the Spanish Civil War, volunteered to join the RAF in a particularly hazardous assignment as a rear-gunner. He was shot down and killed in May 1940.[126] Yeats-Brown supported the British war effort while expressing hope that 'what was good in Nazism will survive'.[127] The rump AGF was shut down, and Domvile ostensibly shut down The Link, though, as will be seen, he did not abandon his activities entirely.

Others chose the opposite course. Pitt-Rivers' old associate William Joyce, who had broken away from the BUF to form the more extreme National Socialist League in 1937, left for Germany before the outbreak of war. He would later become known for his English-language Nazi propaganda broadcasts as 'Lord Haw Haw', and towards the end of the war he attempted to convince British prisoners of war to join a unit called the British Free Corps to fight the Soviets.[128] He would be convicted of high treason and hanged after the war for his actions. Joyce was accompanied in

Germany by Margaret Bothamley, the leader of the Imperial Fascist League with whom Sharpe had attempted to conspire against the AGF's leadership. Bothamley 'was in a state of extreme confusion' in Germany and adorned her flat there with pictures of the Royal Family.[129] Despite becoming a rather piteous figure and being described by a colleague as 'quite helpless', she too made English-language propaganda broadcasts but avoided the hangman's noose at the end of the war.[130]

For his part, Oswald Mosley continued to lead the British Union (as the BUF had been renamed) even after the outbreak of war. Throughout the winter, he held a series of rallies in London and Manchester that attracted hundreds of attendees. However, the party was unable to achieve even small-scale political victories and received only a handful of votes in the three by-elections it contested.[131] More menacingly to MI5, Mosley was among a number of known fascists who attended a series of fortnightly secret meetings after the outbreak of war. The first was held on 26 October and included Mosley, Domvile, Pitt-Rivers, Fuller, Yeats-Brown and other leading fascists and Nazi sympathizers. The discussions at the meeting focused on the creation of a peace campaign and a political strategy to fight by-elections, along with various anti-Semitic aspects.[132] Domvile recorded in his diary that Pitt-Rivers was a frequent attendee of such meetings and also attended a party in Mosley's honour in December.[133]

Like many large estates, Pitt-Rivers' house and village were requisitioned by the military after the outbreak of war. The Hinton St Mary manor house initially served as the headquarters of the King's Royal Rifle Corps (60th Rifles), making Pitt-Rivers effectively a guest in his own home. With his virulent opposition to billeting already public knowledge, Domvile recorded in October that he was 'very funny about the commandeering of his house and village'.[134] Ill-advisedly, Pitt-Rivers showed little hesitation in making his political views known to the officers staying in his home and the nearby village, having reportedly, 'talked to the officers in a violently pro-German manner and impressed upon them his admiration for Hitler, his ideology, his work, the German people and so on' and offering to lecture the troops on similar topics.[135] He allegedly referred to 'my friends Hitler and Goering' and mentioned his connection with Joyce.[136] MI5 passed a warning to Southern Command about Pitt-Rivers' views and expressed concern about the possibility that he might pose a security threat:

> Captain George H. Lane-Fox Pitt-Rivers had been known to us for some time as an adherent of Oswald Mosley, though he is not known to be openly a member of the B.U.F.... He belongs to a small group of persons in this country who are obsessed with the ideas of the dangers of world domination by Jews, international financiers, Bolshevists, etc. While we do not consider him very important politically, there has to be borne in mind the fact that up to the outbreak of war, they maintained close contact with members of the German Nazi Party.... There has to be

borne in mind, therefore, the danger that leakage of information may occur through Pitt-Rivers.[137]

There is no evidence that Pitt-Rivers ever leaked military information, and much of the evidence MI5 was able to gather about his activities related to the guests he hosted at his estate rather than his own actions. Domvile, Fuller and Mosley all visited the house after the start of the war, occasionally attending parties being held for the troops stationed in the house and in the village.

In May 1940, Fuller and his wife, along with Gladys Fortune, attended a dance held for the soldiers in Pitt-Rivers' Tithe Barn. The discussion quickly turned political, and Fortune impoliticly told the wife of a neighbour that: 'If I had a son, I would rather shoot him with my own hands than have him fight against the Germans.' The fact that the couple did indeed have a son who would presumably soon be entering the conflict made the remark sting even more acutely.[138] Pitt-Rivers attempted to explain away the statement by regretting that Fortune's 'real patriotic enthusiasm should have got badly mixed up with her emotional history and her temperamental peculiarities'.[139] Regardless, an account of the incident and the resulting correspondence ended up in the hands of MI5. Still himself obsessed with the idea that the Jews were behind the war, he wrote to Mosley in January 1940 (though he did not send the letter through the post, fearing it would be opened), encouraging him to investigate the activities of the 'Hidden Hand' in various organizations including the Eugenics Society, which he now claimed had been taken over by communists and Jews.[140]

The war was by now taking an ominous turn for the British. In October 1939, the *Wehrmacht* captured Warsaw only weeks after crossing the border, and in April 1940 Hitler launched an invasion of Denmark and Norway in an effort to secure naval bases and resources. British, French Foreign Legion and Polish troops landed in the middle of the month in an effort to defend strategic positions in Norway but were driven out in June. The Germans also managed to sink the aircraft carrier *HMS Glorious* in a surprise attack, killing hundreds of British sailors. The King of Norway and the government evacuated the country, along with the remaining foreign troops who could escape, leaving it in the hands of the German-backed fascist leader Vidkun Quisling.[141] This catastrophe was the end for Neville Chamberlain, who resigned as Prime Minister after forty Conservative MPs voted with the Labour opposition against his government. He was replaced by Winston Churchill, who formed a government of national unity and prepared for the German onslaught against Britain that was looking increasingly likely.[142] Already suffering from terminal cancer, Chamberlain died, humiliated, in November.

There was worse to come. On 10 May, German troops moved into Holland and a bombing raid four days later levelled the centre of Rotterdam at the cost of hundreds of lives. The Germans then moved into Belgium, where King Leopold II surrendered on the 28th. France itself had been invaded through the Ardennes Forest at the beginning of the offensive, and five days later the French Prime Minister told Churchill that his country was defeated. On the 20th, German panzers reached the English Channel, pushing French and British forces towards the port of Dunkirk. Desperate to evacuate as many men as possible, the British assembled hundreds of vessels of all sizes to cross the Channel and carry back as many soldiers as they could. Under heavy fire, more than 300,000 men were rescued, but their equipment was lost. By 14 June, the Germans had entered Paris, and on 22 June, the country officially surrendered and was partitioned between an occupied region in the north and a second state managed from the town of Vichy in the south. Hitler insisted that the surrender be signed in the same railway carriage, placed at the same spot near Compiègne, in which the Germans had signed the 1918 Armistice ending the First World War. On 28 June, Hitler made a visit to Paris and had his photograph taken in front of the Eiffel Tower.[143]

With the surrender of France, Britain now faced the threat of German invasion. The capture of French airbases in the west of the country put much of the country – and London – within range of *Luftwaffe* bombers. It was only a matter of time before the German air raids against mainland Britain began, and it was quite possible that they would serve as the prelude to a full-scale onslaught. Yet there was also something else happening outside the public eye that would have significant ramifications for the British far right and Pitt-Rivers himself. On 20 May, the same day German tanks reached the English Channel, the Security Service raided the flat of Tyler Kent, an American who worked in the US Embassy as a cypher clerk. Kent was found in bed with his mistress, and a search of his flat turned up more than a thousand stolen documents from the Embassy along with the membership list of the Right Club, a secret right-wing organization run by Archibald Maule Ramsay, MP for Peebles. Ramsay had been elected in 1931 and done little to attract attention until 1937, when he began making rabidly anti-Communist and pro-Nazi statements. The following year, he purportedly discovered that communism was a Jewish plot for world domination and fell into the anti-Semitic right. He founded the Right Club in 1939 to 'expose the activities of Organised Jewry' and kept the membership secret to avoid infiltration.[144] The fact that Kent was in possession of the secret membership list suggested a close association with Ramsay.

Worryingly for both the Security Service and the American Embassy, the documents stolen by Kent contained decryptions of a number of cyphered telegrams between Churchill and President Franklin D. Roosevelt. The release of these documents could be dangerous to both leaders given that the United States was still officially neutral in the conflict. Kent had already given key documents to Anna Wolkoff, the daughter of a White Russian naval officer and a member of the Right Club, who had arranged for them to be photographed. She was also arrested. Most dangerous, however, was the fact that Kent had made Ramsay aware of the documents. As a member of Parliament, if Ramsay were to introduce the documents under parliamentary privilege or ask a question about them in the House of Commons nothing could be done to stop him from revealing what he knew.[145] The consequences could be politically devastating on both sides of the Atlantic, particularly with Roosevelt seeking an unprecedented third term in office later that year. Kent, Wolkoff and Ramsay clearly had to be stopped, but what if the plot went deeper? There was no way of knowing how far the Kent documents had circulated and how many of Ramsay's associates – possibly even including Mosley – might be involved.

Faced with this dangerous situation, on 22 May the War Cabinet was informed of Kent's activities and the decision was made to immediately arrest and detain key members of the British Union and other fascists. Ramsay and Mosley would of course have to be among the first detained, as MI5 feared that there might be contingency plans to take the organization underground in the event of repression to prepare for a violent confrontation or even a coup.[146] The same day, parliament approved the Emergency Powers (Defence) Act and the Privy Council passed a modified version of Defence Regulation 18B, a regulation that had previously been used detain enemy aliens and individuals suspected of involvement in spying or other demonstrably disloyal activity.[147] The expanded version, called 18B (1A), gave the Secretary of State the power to order indefinite detention, without recourse to trial, for anyone suspected of being a member of an organization 'subject to foreign influence or control, or (b) the persons in control of the organisation have or have had associations with persons concerned in the government of, or sympathies with the system of government of, any Power with which His Majesty is at war'.[148] In essence, the new version of 18B was written to allow the indefinite detention of anyone believed to have fascist sympathies. The British Union itself would be officially outlawed as an organization by the new Regulations on 10 July.

On the morning of 23 May, Mosley was arrested and detained, along with Ramsay and dozens of leading BUF members. Mosley's wife would be detained the following month. Documents and membership lists were seized from fascist headquarters, giving the Security Service vast insights into the British Union and the wider fascist movement. In early June, several hundred more suspected fascists were picked up.[149] Local constables and Scotland Yard were asked to provide names of individuals in their areas suspected of being British Union members or harbouring fascist views. On 19 June, the Dorset Chief Constable wrote to the Home Office suggesting Pitt-Rivers be detained 'on the grounds of his Fascist and pro-Nazi sympathies'. The letter went on to cite as evidence his published work ('almost in themselves sufficient to justify an Order under D.R. 18B'), his recent houseguests, his pro-German and anti-Semitic statements, his hiring of Tildesley ('a prominent Fascist') as his solicitor and Fortune as his secretary, along with the conduct of his 1935 parliamentary campaign. The Chief Constable continued:

> To sum it all up I would say that, in spite of the fog of rumour and suspicion in which this whole matter is wrapped, I am of the opinion that this man is not only an ardent Fascist but is definitely a believer in and supporter of the Hitler regime.
>
> He is undoubtedly an able man with an excellent brain, but he has a queer warped mentality which might easily herd him into strange paths....Captain Pitt-Rivers has undoubtedly formed powerful contacts in Germany and has the brains and intelligence to use them to the great detriment of this country if he so desired.
>
> If by any chance he is innocent then I would suggest that he has only his own conduct and mode of life to thank for the position in which he now finds himself.[150]

Around midday on 27 June, an inspector from the Dorset Constabulary, accompanied by a number of police officers, arrived at the Hinton St Mary manor house to serve the 18B detention order. Pitt-Rivers had planned to lunch that day with several officers billeted in nearby villages, but one guest instead received orders to mount his Bren machine gun in Pitt-Rivers' front garden with the muzzle pointed towards the front door of the house.[151] When the police arrived at his door, according to an investigator present, Pitt-Rivers 'appeared somewhat taken aback but nevertheless maintained a courteous attitude. He made some remarks about this being a reward for patriotism and referred to his service in the Great War, his literary work etc'.[152] The text of the detention order referred to both his political activities and his writings:

> You, George Henry Lane Fox Pitt-Rivers, were at the date of your detention or had been a member and supporter of the said Organisation [the British Union] and had been active in the furtherance of its objects by contesting the Parliamentary Election as an Independent Anti-Tithe Candidate, by constantly over a period of years expressing verbally or in writing, both publicly and privately pro-German and anti-British views and by distributing Nazi propaganda.[153]

Barry Domvile and his wife were staying in the manor house that morning. They were allowed to leave without incident – only to be arrested themselves on 8 July for their involvement with The Link – and Domvile recorded in his diary that Pitt-Rivers had been taken from the house 'trying to be brave, but looking ghastly – and babbling'.[154] The house was subsequently searched from top to bottom, and Pitt-Rivers' archive of correspondence was seized. Little of interest was found beyond the letters in his study.

Prime Minister Winston Churchill became aware of Pitt-Rivers' arrest a day later when the Home Office presented a list of 'prominent' individuals being detained. It included Pitt-Rivers and Diana Mosley, both of whom were relatives of Churchill's wife Clementine through her Stanley forebears.[155] As a sign of the importance he was assigned by MI5, Pitt-Rivers was fourth on the list, appearing ahead of even William Joyce's family members.[156] Churchill's private secretary, Jock Colville, described the Prime Minister as 'piqued' by the news of Pitt-Rivers' detention. The Churchill children derived amusement and 'merriment' from the news, teasing their parents about their questionable relations.[157]

From Dorset, Pitt-Rivers was conveyed to Brixton Prison in south London. On 2 July, he was visited by an MI5 officer who observed him to be under mental strain:

> His pose was that of a gentle and patient literary man, detained for a reason he did not clearly understand, but ready to assist the authorities with information as he had always done in the past. He seemed anxious to show that he was quite a different person from the other internees, and referred to himself several times as a 'scientist' and a 'political diagnostician'.
>
> It was almost impossible to keep him to the point. I said many times I was not interested in the stories he tried to tell of events twenty years ago, but would be interested in any recent information regarding subversive activities in this country. He appeared a little surprised that a literary man of retiring disposition should be expected to know about these things …
>
> We know that some Fascists have announced their intention of pretending to be mentally unbalanced, and Pitt-Rivers appears to be adopting these tactics. The interview was a complete waste of time.[158]

Pitt-Rivers' mental state was undoubtedly agitated, though not because he was faking madness. As the members of the 18B Advisory Committee reviewing his case would soon recognize, he was legitimately appalled by his circumstances and could not understand the reason for his detention. His insistence on being treated as a 'scientist' was not obfuscation but was the way he genuinely viewed his activities.

Pitt-Rivers was subsequently transferred to Liverpool Prison and then to the winter quarters of Bertram Mills's Circus at Ascot, which had been converted into a prison camp for more than 600 detainees. Conditions at Ascot were difficult, and the detainees slept on wooden bunks that aggravated Pitt-Rivers' war wound and resultant arthritis.[159] Food was in short supply, and the Ascot kitchen was made to provide nutrition for the entire camp with rations intended for just 400 people.[160] Liverpool Prison had been an additionally trying experience, as he told the Home Office in a petition for release:

> Bugs and vermin abounded: the water-closet on our ground floor Block was choked for days. The stench of faeces and urine which flooded the sanitary well made me think longingly of the nice clean dug-outs on Vimy Ridge over twenty years ago. Weeks of 22 hours a day solitary confinement, denial of my pneumatic mattress, and a bed board for a crippled leg and arthritic hip are punishments enough for my present application for release on medical grounds.[161]

In mid-September, he penned a scathing letter to his cousin, Clementine Churchill. Addressed to Mrs. Winston Churchill, 10 Downing Street, the note was loaded with acerbic sarcasm:

> My dear cousin Clementine,
> How are you and Winston, and all the family? I was glad to hear about Randolph and the little Digby girl, I hope they are very happy.[162] The last time I saw him at Quaglino's [restaurant] he was looking very well and very smart in his nice 4th Hussar uniform …
> As for us – we are, of course, being very well looked after and the camp officers are very kind. We have divine service every Sunday and we sing God Save the King at attention. There are tears in many eyes.
> Do you remember what Shakespeare said about '5th Columnists'?
> 'O let me have no subject enemies, when adverse foreigners affright my towns with dreadful pomp of stout invasion!'[163]
> Winston might well quote in his next speech.
> Give love to cousin Sylvia.
> Your affectionate cousin,
> George Pitt-Rivers
> Camp No. 1350
> P.S. In case you have left 10, Downing Street, I hope the butler will forward. I am only allowed to write 24 lines.[164]

There is no evidence Clementine or her husband ever saw the letter, and it does not appear to survive in the Churchill family papers. Pitt-Rivers, however, seems to have derived pleasure from the very fact that it had been sent.

Other old friends were more open in their support. Putting their political differences aside, Oscar Levy wrote to Pitt-Rivers just days after his arrest, comparing his current plight to Levy's own difficulties two decades before:

> I remember a time in my own life, when I too was deprived of my home, and when you did everything in your power to help me over my

difficulties. Your effort was in vain – but that was not your fault. I also remember meeting later in Paris, when you said to me 'I suffer from your misfortune, as if it had happened to myself.' It was generous and I must and cannot forget it, even if of late we may not have seen eye to eye in politics. Politics pass – friendship should remain. It is in memory of this friendship and in agreeable meetings in the past, that I write you these lines.[165]

In a reversal of their previous roles, Levy wrote to the Home Secretary on Pitt-Rivers' behalf, sent his daughter to visit him in Brixon Prison and attempted to arrange accommodation for him in Oxford in the event that he were released. Their correspondence retained a philosophical tone. 'Was it perhaps, that you, as an honest scientist, were too "objective"? It is dangerous now days to be so – I know it from my own experience, though my own objectivity was only applied in the realm of philosophy. Even here it is not without peril', Levy wrote.[166] Writing from his prison cell in Liverpool, Pitt-Rivers told Levy that their views were not as far apart as their past conflict had suggested. 'The only difference between us, I conceive, is that I get much too serious and excited about philology and the abuse of words and the silly arguments about the pedigree of Shem [the origin of Semitic peoples], about which men will fight to the death, because they think they are fighting about something quite different, whilst you are just content to laugh at them', he told Levy.[167]

In November, Pitt-Rivers was called to an advisory board panel tasked with examining the case for his detention. In a show of political protest, he refused to report to the camp gates as ordered. A small riot broke out among the other prisoners when soldiers removed him by force with bayonets fixed on their rifles.[168] For the Security Service, the episode was a stark illustration of the danger Pitt-Rivers might pose by instigating trouble if released from custody, particularly if he returned to an unstable part of the country.[169] For Pitt-Rivers himself, the incident was another demonstration of how far his country had fallen. 'It is of small consequence that this "charge" took place with fixed bayonets', he wrote later, 'since it was commonly believed that bayonets were never unfixed because the sockets had rusted on to the muzzles of the rifles.'[170] Rusted or not, the rifles and bayonets would keep Pitt-Rivers in detention for months to come as the British government deliberated the threat – or lack thereof – that he posed to national security. As the 18B Advisory Committee would soon itself conclude, Pitt-Rivers presented a very different case from the average detainee.

CHAPTER SEVEN

A World Destroyed, Again

Being detained under Defence Regulation 18B during the Second World War was akin to falling into a legal black hole. The normal rules of the British justice system no longer applied: there was no requirement on the government to demonstrate the causes for arrest and internment (*habeas corpus*), and there were no regular legal hearings or trials in which the accused could protest their innocence or present evidence. There was no opportunity for the accused to examine the evidence against them or, in many cases, even know the identities of their accusers or the claims that had been levelled against them. Without trials, there was no way to actually be convicted or acquitted – detainees simply remained in custody until the government decided to release them. As A.W. Brian Simpson remarked in his famous study of the 18B detention system, over the course of a few weeks Britain had 'become, in the name of liberty, a totalitarian state'.[1]

Pitt-Rivers later evoked the twisted logic of Lewis Carroll's *Through the Looking-Glass* (the sequel to *Alice in Wonderland*) to lampoon the logic at work:

> By an ingenious 'Looking-Glass' logic the ancient constitutional safe-guards [of *habeas corpus*, etc.] were by-passed: by making suspicion of harbouring an undisclosed intention a '*reasonable cause*' to imprison a suspect. It was therefore not prudent or possible to make any accusations. The principle was explained by the succeeding Home Secretary ... in terms which appear to rely on the precedent established in 'Alice through the Looking Glass'.
>
> 'It is not what they have done which matters, it is what they may do, if we do not prevent them from doing it before they do anything,' said the Minister of the Crown.[2]

There was little protest against 18B among the public, and a Mass Observation report filed in late May 1940 reported that the arrests and detentions were very popular and in line with the generally negative mood towards the fascists they targeted.[3] By the end of the war, 1,847 detention orders had been served by the British government, the majority justified by the alleged 'hostile origin and/or associations' of the suspect.[4]

No one at the Home Office or MI5 ever seriously argued that Pitt-Rivers posed a direct threat to national security. There was no evidence that had ever passed sensitive information to the Germans, and, indeed, he appears to have corresponded very little with his German contacts after early 1939. The outlandish stories circulating in Dorset about giant arrows pointing towards London being planted in the hedges and machine gun nests in his front garden had been found to be without substance by the authorities who raided his home. Even the search of Pitt-Rivers' home study and office had revealed little recent material of use to the Security Service beyond a few letters to fellow fascists. These could not have revealed much useful intelligence to MI5 since virtually all the recipients were already under surveillance themselves and many were already interned.

The most plausible case that could be made for Pitt-Rivers' detention was that that he was a nuisance and might conceivably damage the war effort through his public statements and writings. By early 1941, even MI5 conceded that he posed virtually no threat to national security beyond publicly expressing his views, which might harm morale, but was 'a man of weak character and ready to become the tool of any fanatical Fascist, whom he comes in contact with'.[5] The Governor of Brixton Prison reported simply that he was 'contentious and argumentative, and likely to be a nuisance wherever he is'.[6]

The only legal route available to an 18B detainee was a written appeal to the Home Secretary and appearance before the 18B Advisory Committee. The committee was composed of well-respected barristers who reviewed cases and made non-binding recommendations to the Home Office about what should be done with the detained. Its two chairmen were Sir Walter Monckton, a former adviser to Edward VIII who also worked for the Ministry of Information and never actually held a hearing, and Norman Birkett KC (later to become Baron Birkett), a former MP and leading barrister who would serve as an alternative British judge at the Nuremberg Trials.[7] For the Pitt-Rivers hearing, Birkett was joined by Sir Arthur Hazelrigg, a baronet who was thought to 'know the German mentality'; Sir George Clerk, a retired Ambassador; and G.P. Churchill, another retired diplomat who served as the committee secretary.[8]

Utilizing the substantial financial means at his disposal, Pitt-Rivers hired two solicitors to represent his case. The first, Claremont Haynes & Co., was Richard Tildesley's firm and was viewed with suspicion by the Home Office because of his presumed fascist views (Tildesley himself was eventually called before the advisory committee on the suspicion that Pitt-Rivers was leaking confidential information through him) and the second, Wainwright & Co., appears to have been more effective in presenting his case. By July 1940, Pitt-Rivers' legal representatives were already protesting that his living conditions were taking a significant toll on his health:

> The conditions under which he is being detained – involving long periods of cellular confinement – are naturally not only distressing, but we submit actually harmful to a man of our client's physical and emotional temperament, so much so that his family are of opinion that any long continuance of the present detention is likely to exercise a permanent prejudicial effect upon our client's health and condition.[9]

The camp medical officer disagreed, reporting that his health 'is not adversely affected, physically or mentally, by detention. There is no need for his release'.[10] In October, Pitt-Rivers made his first appearance before the committee of Birkett, Hazelrigg and Churchill in a short hearing that was quickly adjourned until the following month. The only outcome was that Pitt-Rivers requested the ability to freely consult his solicitors, obtain paper and writing implements and be allowed to see his son Michael.[11] The hearing ended without a commitment from the committee that any of these requests could be met, though he was subsequently given access to legal counsel.

In early November, Pitt-Rivers was once again called before the committee for a more substantial hearing that would in part determine whether he could be released from detention. Pitt-Rivers' attitude towards Birkett was combative, and the discussion immediately focused on the legal basis for his detention. Under the Defence Regulation 18B, the only grounds that could be used to justify detention hinged on him having been a member of the British Union. This allegation he strenuously, and truthfully, denied. His involvement with Mosley during the 1935 parliamentary campaign was a purely strategic move, he claimed, and 'I did not regard myself as a member of the British Union, and to the best of my knowledge, Oswald Mosley, who should be capable of determining it, never regarded me as a member of the British Union.'[12] Birkett questioned the BUF's support for his candidacy, and Pitt-Rivers simply deflected the question by arguing that he had also been supported by Labour Party members. In actuality, he claimed, he was never a politician at all, as the following exchange with Birkett made clear:

> Q [Birkett]: … I understand you to say this, that your connection, if any, with British Union was rather loose and casual, never regarded by you as being an enrolled membership, and any association was entirely due to the fact that you had an association for certain objects of your own and British Union found it politic in some measure to support your policy. Does that put it right?

A [Pitt-Rivers]: Well, that is as you put it and not as I put it.

Q: I do not want to put it other than how you would put it. How would you put it?

A: I would have put it this way: we do not want a short answer. I would have to make it perfectly clear, as I intended to do at the beginning, that I look upon myself, and indeed I claim to not be a politician but a man of science, and a student. That also means, as an anthropologist, an ethnographer and a sociologist, that I am a student of politics, and I would like you, if you will, to distinguish between an expert student of politics and a politician.

Q: What is an ethnographer, an anthropologist and a sociologist doing on a political platform as a candidate?

A: That is the point I wish to make.[13]

For hour after hour, Pitt-Rivers tied his questioners into rhetorical knots with this type of logic, repeatedly turning the tables on the questioners. Birkett himself was a former MP, and Pitt-Rivers told him, 'you did not cease to become an expert jurist or man of law because you either tried to become or became a Member of Parliament, but far less, whether I had succeeded or not, should I have become a politician then [sic] you became one'.[14] Asked whether *The Czech Conspiracy* was pro-German, Pitt-Rivers retorted that with its writing he had simply been, 'offering my training and my expert knowledge to the Government at the time', and he told the committee that his opposition to billeting had ended once the war broke out.[15] His contacts in Germany had been made through his involvement with various scientific organizations including the IUSIPP, he told them, and the fact that some of those individuals had later taken positions in the German government was pure coincidence. The committee naturally raised the question of the anti-billeting petition that had been circulated around his estate, leading him to lose his temper and spend more than a page of the transcript denouncing John Coast for his extremism and his relationship with Becky Sharpe.

By the end of the day, the hearing had made little progress. Pitt-Rivers brushed away the suggestion that he had been unduly sympathetic to Hitler, insisting that he was merely a scientific practitioner and student of politics who had tried to offer his expert services to the British government. His appearance at the 1937 Nuremberg Rally and all his connections in Germany were 'in connection with my University lectures arising out of my work as General Secretary to the International Conferences'.[16] Most of his German friends, including Eugen Fischer, were objective scientists who had been appointed to governmental posts after Hitler's ascension to power in 1933, he claimed, and could hardly be described as Nazi ideologues. He admitted to meeting Hitler in 1937 but 'never exchanged a word with him' and denied ever referring to the *Führer* or Goering as his friends. Ribbentrop he had only met 'once or twice'.[17] He denied having written to Hitler to congratulate Germany on the *Anschluss* but when pressed by Birkett, who had of course seen the letter seized from his study, admitted that he might have done so, though 'I do not recollect such a letter.'[18] Exhausted by these rhetorical circles, the Committee adjourned after hours of questioning.

Three days later, Pitt-Rivers once again appeared before the panel. It was 11 November – Armistice Day – and he began by once again turning the tables on his questioners in a passionate monologue:

> On the 11th November, 1918, I was at the front, slightly crippled even then as the result of wounds received in the Ypres salient. My two sons are today in the King's uniform fighting as I was then fighting for King and country. For 22 years I have served my country in science, agriculture and for my country folk and laboured that they and my own sons, one not yet born, should not go through the hell we went through then. I have not during all that time since laboured for place or position; may be if I had I would not be where I am today, and today I stand before you, not in the King's uniform but straight from the King's prison. How odd, I ask you, Mr. Birkett, if the Unknown Warrior beneath the Cenotaph had lived instead to be brought to you on Armistice Day a prisoner as I am, and yet the King's trusted and faithful friend on two King's commissioning parchments.... On the subject of peace and war I have through all these years of precarious peace thought as a soldier. It is not soldiers who make wars; politicians make wars.[19]

Birkett was now better prepared to question Pitt-Rivers than he had been the previous week. He now carried copies of all his correspondence with Mosley, seized from Hinton St Mary and passed to him by MI5 during the weekend between the hearings. Once again, the enquiry hinged on Pitt-Rivers' relationship with the BUF upon which the detention order had been predicated. Pitt-Rivers again denied being a member of the party or a politician in any sense. His contacts with Mosley, he told the committee, were born out of efforts to organize against the tithe system. 'Mosley has never regarded me as an adherent of his British Union,' he insisted.[20] Questioned about his allegedly pro-German remarks to the officers visiting his home, Pitt-Rivers admitted that he might have made unwise statements, but was merely offering opinions as an expert on Germany and military strategy generally.[21] Everything he had done, written, and said was based in his scientific work, he claimed, and was by its nature non-political.

Birkett and his colleagues were grudgingly sympathetic to Pitt-Rivers' plight. Two weeks later, G.P. Churchill filed the official report of their conclusions on 'this unhappy case'.[22] 'This Committee has had many difficult people to deal with and difficult decisions to make, but none of those who have appeared before them have occasioned them so much trouble as Captain Pitt-Rivers,' he wrote. Pitt-Rivers was motivated by 'what, to his mind, have been motives of the most sincere patriotism', and was 'a man of high intellectual attainments, but filled with intellectual conceit and warped by the lack of recognition with which his national and international work has been met by the various Government departments to whom he has expounded his views'. In addition, Pitt-Rivers 'is a man who is constitutionally incapable of giving a direct answer to a simple question' and a 'hair-splitter' who 'takes refuge behind a cloud of scientific or philosophical verbiage'.[23]

Significantly, the committee concluded that there was no basis for the existing detention order that had been issued for him, because 'in no real sense of the term was Captain Pitt-Rivers a member of British Union'. However, it did agree in principle that Pitt-Rivers should continue to be detained because releasing him would allow him to 'give utterance to his views which, in the opinion of the Committee, would not only cause trouble locally, but would have the effect of weakening the national effort'.[24] Because continued detention might take a negative toll his health, 'the Committee would therefore urge that his detention should be made as little rigorous as possible, preferably somewhere where his health could be properly looked after and where he could prosecute his scientific studies'.[25] Concessions were thus granted for Pitt-Rivers to purchase writing utensils, paper, and books at his own expense.[26]

As much as the law still mattered under these circumstances, the government was now in somewhat of a legal quandary. The committee had effectively advised the Home Office that the legal basis for Pitt-Rivers' internment was unjustifiable. He had no meaningful association with Mosley or the BUF, and he had convinced the members that his connections in Germany had been legitimate scientific ones. As the Home Office itself concluded, the only danger he presented was in the fact that, 'he is quite incapable of controlling his tongue and that his views might well have a subversive and harmful effect and very likely arouse fierce antagonism'.[27] One possible solution was issuing him with another detention order based on the 'acts prejudicial' he might commit, but the Home Office rejected this course on the grounds that it might encounter 'formidable difficulties' making the case to secure such an order.[28] By early 1941, the government was looking for a way out of this bind. A Home Office minute noted that the best solution would probably be to send Pitt-Rivers to a nursing home where he could undergo treatment for his war wounds while being subject to restrictions on his movement and 'which enabled the police to keep an eye upon him'.[29] The one caveat was that he could not be allowed to return to Dorset.

Pitt-Rivers himself now proposed a solution to the committee. Writing through his solicitor, he asked to be released to a nursing home in either Oxford or London to undergo treatment for his old war wound and associated arthritis. If allowed to live in Oxford, he asked to reside in Worcester College 'in order that I may continue my Agricultural Research and Archaeological Research Work, both practical and theoretical – in which I am greatly interested'. He pledged to 'refrain from any form of public or political activity whatsoever' and additionally promised not to associate 'with persons who by reason of relationship or of political associations may be objects of suspicion'. Beyond his 'honourable undertaking as an Officer and a Gentleman for the due performance of these terms', he offered as sureties his son Michael, his former wife Rachel, and his solicitor Richard Tildesley.[30] The advisory committee was reasonably convinced by these offers and believed they might offer a face-saving solution for the government. 'If all these terms are obeyed, then the advisory committee feel that all the objections discussed by them in their original report are resolved and that the release of Captain Pitt-Rivers under these restrictions is the best solution of this unhappy case,' G.P. Churchill told the Home Office.[31]

The Security Service now intervened. Four days after the committee seemed to embrace the idea of a conditional release, MI5 formally objected on the grounds that 'he is a man who is unable to refrain from talking' and might cause trouble anyplace he ended up.[32] His promise to abstain from associating with other people under suspicion would be impossible to enforce, MI5 feared, because Pitt-Rivers did not possess a list of such people and it would obviously be undesirable to provide him with one. In addition, Tildesley would not be a reliable surety because he himself was under suspicion, Michael could not act as such because he was serving with the Welsh Guards, and his former wife, Rachel, would not be living with him. Finally, sending him to Oxford was completely out of the question because, 'it is undesirable that he should be in a position to give vent to his objectionable views in the University' and, tellingly, because 'he would be entitled to wander about Oxford and might meet some of our officers who interrogated him'.[33]

The plan to release Pitt-Rivers was thus quashed for the moment, and the Home Office demurred in formally making any kind of decision until late August.[34] In the meantime, Pitt-Rivers remained in Brixton Prison. By September 1941 he was on crutches from the arthritis in his wounded left leg. He complained to a doctor examining him that he was 'always completely paralyzed in the morning. I am in constant pain. I get pain in bed. I am not sleeping well.'[35] The doctor advised that he be moved from Brixton to obtain proper treatment, with another reporting that his hip was showing signs of significant change as the result of osteoarthritis.[36] The prison medical officer disagreed, reporting instead that Pitt-Rivers was 'a very difficult man' and used crutches when there was a 'sympathy parade' but normally used only a walking stick. Detention had 'in no way contributed' to the progression of his arthritis, in this doctor's opinion.[37]

Frustrated by the delay, Pitt-Rivers penned a scathing and sarcastic letter to the Home Secretary that was probably never sent but reflected his increasing anger:

> I now fully appreciate and am duly contrite and repentant for my past errors and sins, being a faithful and loyal Christian and therefore know how vile a sinner I am, full of error and rebelliousness, yet very humble and conscious of my wickedness. I am now very ready to make amends and acknowledge the true Christianity which joins our Holy Cause with that that of the Russian Soviet government, and with your own superior wisdom and goodness. I am opposed to all vile Nazi propaganda and freely acknowledge that I have in the past been deceived by it....Puffed up with conceit and ignorance I acknowledge that all my published scientific work is only folly...and that only H.M. Government and his Grace the Archbishop of Canterbury are authorities in these matters. If I declare my loyal belief in any or all that H.M.'s Ministers declare for the time being to be true...may my case be reviewed with a view to release? I believe in V for Victory, dot, dot, dot dash.[38]

Regardless of how many angry letters Pitt-Rivers could bring himself to pen, he was once again stuck in legal limbo with the Home Office refusing to take action in the face of MI5's objections. He now had only one legal recourse remaining: filing a writ of *habeas corpus* in the Royal Court of Justice. Under normal circumstances, *habeas corpus* would require the government to bring a prisoner before the court to present charges and evidence against them. However, the Law Lords would soon dismiss several similar writs by other 18B detainees and a lawsuit for false imprisonment (the famous case of *Liversidge v. Anderson*). In all cases, they ruled that the Home Secretary's belief alone that a detainee was a threat to national security and the war effort was itself sufficient cause for continued detention. These rulings effectively granted the government unlimited executive power to hold anyone without charge for the duration of the war.[39] The Home Secretary was not even required to submit an affidavit stating his views about the dangers the subject posed: the existence of the detention order itself was ruled to be reasonable cause to exercise control over the detainee. Further, the Home Secretary could not be asked to show the sources of his evidence against the detainee due to the sensitivity of MI5's sources.

Pitt-Rivers filed his suit in late August 1941, weeks before the *Liversidge v. Anderson* ruling. His affidavit argued that the original detention order was invalid because, firstly, it contained no explicit written statement that the Home Secretary believed it was necessary to exercise control over him and, secondly, because it was predicated on his membership in the British Union, which had been dismissed by the advisory committee.[40] MI5 provided its pre-detention outline of the case against Pitt-Rivers to the

Lords considering the case, including the fact that he had been known to the Security Service since 1930 'as a strong anti-semite [*sic*] and an associate of those who share his views' and had contacts in Germany. His detention, it concluded, 'is a matter of urgency'.[41] It included copies of intercepted letters written to Mosley, including a June 1934 letter in which Pitt-Rivers reported joining the pro-fascist January Club and 'procuring some other sound members'.[42]

In December, the Law Lords rejected both of Pitt-Rivers' arguments. There was no need for the initial detention order to contain an explicit recitation of the Home Secretary's beliefs, they argued, because 'the fact of his detention was the plainest intimation to him that the Secretary of State considered it necessary to exercise control over him'.[43] Secondly, because there were no grounds to question the veracity of the Home Secretary's views, the factual status of Pitt-Rivers' membership (or lack thereof) in the BUF was less significant than the Home Secretary's belief that control had to be exercised over him. Finally, they ruled, even if 18B had exceeded the traditional interpretation of *habeas corpus*, 'what parliament has done parliament can undo when it thinks fit'. In other words, because parliament had passed the defence regulations, the court was unwilling to intervene against them in a time of war.[44]

Ironically, by the time Pitt-Rivers' *habeas corpus* writ was rejected, he did not have long to wait for release. Throughout the summer of 1941, most of the 18B detainees were transferred to the Isle of Man with only a few, including Mosley, Pitt-Rivers and various BUF leaders whose influence on the other detainees might lead to mass defiance, remaining in Brixton. Two dozen women, including Lady Domvile and Lady Mosley, remained in Holloway Prison.[45] Politically, Winston Churchill was now concerned about the effect 18B might have on public opinion at home and abroad, particularly in the United States. It was hardly desirable to have to defend the repression of unpopular views that did not demonstrably pose a threat to national security without trial or legal recourse when the country was fighting a war against fascist tyranny. Detainees whose health was endangered by continued detention were therefore seen as particularly important to release after the initial German invasion scare had passed, and in November 1943 a case of phlebitis was seen as sufficient to release Mosley himself.[46] The same month, Churchill circulated a telegram to the Cabinet arguing that 18B should be terminated as soon as possible as it was 'contrary to the whole spirit of British public life and British history'.[47]

Pitt-Rivers could make a particularly compelling case that his health was endangered by his continued detention. His doctors assiduously filed reports stating that his condition was worsening and that he required an immediate operation to avoid long-term damage to his health. The Security Service had already rejected the idea of him going to Oxford, as he had proposed, but a compromise would be to release him to undergo a procedure and recuperate elsewhere.[48] In early December 1941, he was sent from Brixton to a clinic in London to undergo an urgent operation on his leg. On 16 January 1942, the Home Secretary, Herbert Morrison, modified Pitt-Rivers' detention order to release him from Brixton on the grounds that he must stay with his sister in Tonbridge Wells, Kent, unless permission were specifically granted for him to live or travel elsewhere. Dorset and Oxford were both forbidden, and travel to London was only permitted for medical consultations. If these conditions were violated, he could be returned to Brixton at any time.[49] MI5 immediately ordered his mail and telegrams retained by the post office for review by its agents, and he was placed on the Suspect List of people to be closely watched.[50]

Two of Pitt-Rivers' oldest supporters, Arthur Keith and Oscar Levy, were both elated by the news. 'I am…more glad than I can express at your release by Govt. I'm altogether with you that we cannot study Germany too closely or too attentively…and copy what she has than what we ourselves have arrived it', Keith told him.[51] If Pitt-Rivers needed 'references as to the value of your scientific work', he continued, 'call me as a witness'.[52] He later wrote of Pitt-Rivers' detention in his own autobiography, stating that, 'in my opinion his liberty should have never been suspended; he has, and had, an abiding love for his country'.[53] As he would for Reginald Ruggles Gates, Keith would remain a supporter of Pitt-Rivers until his death.

For his part, Levy remarked that the experience of detention 'must appear like a nightmare to you now'.[54] The previous September, he had taken his daughter to the Pitt-Rivers Museum in Oxford where he paused to consider the fate of the founder's grandson from the perspective of General Pitt-Rivers' work and legacy. 'His grandson, whatever people may say, has remained faithful to the tradition [of General Pitt-Rivers], and has continued and will continue to do good work – in spite of the unwelcome interruption of to-day', he told Pitt-Rivers.[55] He recalled once seeing Pitt-Rivers' wound, which had still not healed years after he sustained it:

> I well remember the wound, which crippled you in 1914 in the fight for England. It was on the left shin, just below the knee. I happened to visit you in your little hotel…on the left bank of the Seine….The wound had not yet healed: it was still painful, I remember, and, when coming to the crowded streets of Paris, you needed to take my arm.[56]

Just as Pitt-Rivers had been unable to prevent Levy's removal from the country two decades before, Levy had been unable to help his friend secure freedom in any meaningful sense. However, the moral support he had offered Pitt-Rivers had been gratefully welcomed. In June 1942, Pitt-Rivers received permission to visit Levy in Oxford. During the visit, Levy urged him to forgive his enemies rather than seek vengeance. 'The best revenge would be no revenge, or better, a fine revenge, an un-Christian, an aristocratic revenge', Levy told him. 'Prove to them that you have profited from your internment, that they, in their injustice, have rendered you a service.'[57]

Release from Brixton and the suspension of the detention order did not mean full freedom, and Pitt-Rivers was still subject to the control of the Home Office even after leaving the prison gates. By March, he was recovering from a procedure on his leg and potentially facing a return to the prison if the Home Secretary deemed him to still be a threat to national security. Birkett was sympathetic to Pitt-Rivers' plight, offering to bring his case forward for consideration, 'for I think he has been harshly treated'.[58] Again appearing before the advisory committee in late March, Pitt-Rivers was advised by his solicitors that he would likely be released without conditions, 'unless during the hearing you say something about your intentions or views which alarms them into thinking that you intend to inconvenience and upset the Government if you have the chance to do so'.[59]

The transcript of the second hearing and the supplementary documents that accompanied it do not make for pleasant reading. On 23 April, the advisory committee filed its final report on Pitt-Rivers, again noting that his had been 'one of the most difficult cases with which they have ever had to deal'. The document went on to note that the only way to properly understand the case was to recognize that Pitt-Rivers 'is in no sense of the word a normal man. In colloquial speech he could be described quite truthfully as being "touched"'. In regard to his evasiveness and inveterate hair-splitting, the committee reiterated that these were not efforts to conceal malfeasance as 'Pitt-Rivers is horrified and shocked that any action of his should be so regarded.' The difficulty lay again in the fact that he seemed unable to control his tongue. In the course of the hearing itself, he casually mentioned that he had encountered and spoken to J.F.C. Fuller that very day, telling the general that he had no idea why he had been detained in the first place. Given that Pitt-Rivers had agreed to forego any contact with other 'suspect persons', this admission was surprising to the committee. In the end, the committee considered simply re-interning Pitt-Rivers to avoid the possible trouble he might cause but agreed that this would probably not be a 'just or proper solution'.[60]

More troubling to all involved was the subsequent intervention in the case by Pitt-Rivers' sister, Marcia Ruth Astley-Corbett. Under the original suspension order, Pitt-Rivers was required to live with his sister, who had been widowed in 1937 and was now tasked with keeping on eye on him. On 16 April, Astley-Corbett secretly appeared before the advisory committee to present evidence about her brother's 'mental condition'. Following his release from Brixton Prison, she told the committee, her brother's 'disillusionment of life altogether, his repeated failures in his domestic and his public life, seem to overwhelm him every now and again, and he has these attacks when he sort of just holds forth and shouts…. He thinks he does not get enough sympathy and understanding, he feels he is misunderstood, and the whole thing is an injustice, and he cannot understand, and just that I cannot understand it seems to be particularly irritating'. Sending him back to Brixton, she said, would 'just finish him' and she plaintively begged the committee to give her some kind of medical support to help him. She complained that he was seeking the companionship of a new female secretary who he might try to make his wife, 'but for the family we really cannot have any more wives'. He had, she said, 'always desired to educate and mould the ideal wife, which is a thing he has projected in his own mind…it is all very complicated'.[61] Understandably reluctant to tread into the inner recesses of Pitt-Rivers' personal life, the committee simply recommended that he remain with his sister and be given access to any medical care that he required. Visits to Oxford and Hinton St Mary were now tentatively permitted on the grounds that they might be beneficial to his overall state of mind.[62]

The experience of detention and the associated accusations of disloyal conduct had obviously taken a severe mental and physical toll on Pitt-Rivers by 1942. The advisory committee was understandably at a loss in deciding what to do with him given the overall situation, but eventually concluded that it should simply suggest the continued suspension of the detention order with his sister as the surety. The Home Office accepted this judgement, and by mid-1942 Pitt-Rivers was permitted to travel with advance permission and receive a limited number of visitors.[63]

By this time, the Hinton St Mary estate had been under government requisition for years. Author Malcolm Muggeridge visited the manor during its occupation by a motor reconnaissance unit and abruptly realized that he had previously worked with Pitt-Rivers' former wife, Rachel, in the theatre:

> They were nearly all young, eager officers, and I greatly enjoyed my time with them. The house where the unit was stationed belonged to G.H.L. Fox Pitt-Rivers who had been interned under Regulation 18b for his pro-Hitler attitudes, and in his library there was an elegantly bound presentation copy of *Mein Kampf,* which was occasionally taken out and looked at, almost reverentially; then put back, without ever being damaged or mutilated. I found this bizarre, but rather endearing. The house was in excellent condition, Pitt-Rivers being, as I gathered, very rich. Despite wartime stringencies, the extensive grounds were all maintained, and there was a neat little theater which Pitt-Rivers had constructed for his wife, an actress. Someone told me that she used as her stage-name, Mary Hinton, after the place, and then I remembered with a pang that she had taken part in my play, *Three Flats*, at its Stage Society production at the Prince of Wales Theatre in 1931.[64]

In addition to the military presence, the manor's Tithe Barn and stables had been converted into a school for evacuee children, straining the facility's electrical supply and plumbing. Pitt-Rivers was worried about the contents of the house and his extensive library, but few concessions were granted for him to visit or obtain his possessions.[65] Unable to return to Hinton St Mary under the terms of his detention order, he found accommodation in a formerly derelict mill house in Uckfield, Sussex, where, he complained to the Director of Quartering, he had water running above the floorboards in winter, and, 'Unless I am assisted instead of further penalised, I know of no other alternative but to retire to my College at Oxford and write a treatise on the "Decline and Fall of the British Empire" or the "Collapse of British Agriculture," perhaps under one heading, "From World War to World Famine."'[66] The Security Service had of course already put this out of the question, but it is telling that he still viewed it as a possibility at this point.

Now fully released from the physical constraints of detention but still obviously under MI5 surveillance, Pitt-Rivers purported to make himself useful to the war effort. Before being arrested, he had begun to reclaim 300 acres of wasteland on his estate for agricultural purposes and he now raised fowl at his house in Sussex.[67] In private, he remained openly contemptuous of the British government and embraced his 18B detention as an indicator of his importance and the sacrifices he was willing to make for 'the truth'. Writing to his aged uncle, the decorated soldier Ulric Thynne, he described MI5 as 'our Gestapo' and acidly claimed that he had been 'the victim of the lie-slander and gibe campaign that mental deficients, who are used in politics, mistake for wit'.[68] Thynne was unimpressed, telling him 'you must admit you asked for trouble and unfortunately the authorities took you seriously and you got it. Those, who *did* know you, of course never took you seriously and would vouch for your harmlessness'.[69]

Despite Pitt-Rivers' continued statements of this type, all of which were known to MI5 as its agents were reading his mail, naval intelligence invited him to provide maps and other intelligence information about the Coral Sea and the South Pacific.[70] He agreed to do so after regaining access to his papers and library. Ironically, one of the officials who contacted Pitt-Rivers for his help was his former PhD examinee Raymond Firth, who was now working for naval intelligence and compiling a handbook on the South Pacific. 'It's been ages since we met', Firth understatedly noted in 1942.[71]

In a further irony, former Home Secretary Sir John Anderson – the minister who had signed Pitt-Rivers' detention order, now serving as Chancellor of the Exchequer since 1943 – moved to a nearby house in the same part of Sussex. Lady Anderson subsequently visited her neighbour to purchase goose eggs and, informed of his feelings towards her husband, ill-advisedly remarked that Sir John had been 'completely unaware of the identity of the persons it was the duty of his Department to advise him to have "reasonable cause" to intern and detain'.[72] This was no doubt the last time the Andersons purchased eggs from Pitt-Rivers: awkward and accusatory letters inevitably followed (in one of which he sarcastically signed his name above the post nominal titles 'B.Sc. (Oxon.), F[ellow of the] R[oyal]. A[nthropological] I[ntitute], egg merchant etc.'), and the eggs themselves turned out to be addled.[73]

As increasing numbers of 18B detainees were released after 1942, Pitt-Rivers remained in contact with those he had known before the war and some he met in Brixton. Elwin Wright, the extremist former secretary of the Anglo-German Fellowship, was predictably detained, and, after being released, stayed briefly with Pitt-Rivers. The irascible Wright evidently overstayed his welcome and offended his host with tales of heroic self-sacrifice, and Pitt-Rivers reported that, 'I became painfully aware, during the latter part of his self-invited sojourn with me that, like quite a number of other people, who like now to wear the crown of 18B martyrdom, he was capable of slandering and working against those he professed to be friends with, and to work for, so long as he could derive any profit out of it.'[74]

Other connections he evidently maintained on the far right were seen as more dangerous. Throughout the Second World War, MI5 operated an impressive intelligence operation designed to gather information about potentially disloyal Britons. An agent operating under the pseudonym 'Jack King' – his real name was Eric Arthur Roberts, and before the war he had worked at the Euston Road branch of Westminster Bank – spent most of the war posing remarkably well as a German agent 'having facilities for communication with Germany and entrusted by the Gestapo with the task of identifying Nazi sympathizers in this country'.[75] In an extraordinary feat of spy craft, Roberts managed to convince six people to act as his 'agents' and report information about fellow Britons holding fascist and pro-German views for what they believed would be future missions to help the German war effort. In reality, Roberts was reporting everything he learned to his superiors at MI5 and the flat where he met with his recruits was evidently bugged.

As the war drew to a close, Pitt-Rivers was listed among the 'Hitler worshippers' about whom information had been received through Roberts' operation. As the Security Service itself noted, this did not mean that Pitt-Rivers was intending to act as a German agent – though the six people directly recruited by Roberts, most of whom were women, certainly believed they were acting on Hitler's orders and some even expressed disappointment that British civilian casualties were not higher – but merely that the agents had provided information about his activities and views that they believed might be useful to the German side at some future time. The fact that his name appeared on the fairly short list of key people about whom information was obtained suggests that he maintained extensive contacts among the far right and continued to express views that attracted attention.[76]

In Pitt-Rivers' view, the war was a disaster of shocking proportions. Even after its end, and the revelations of German atrocities across Europe, he showed no hesitation in claiming that it should never have been fought, perhaps confirming MI5's earlier supposition that he would be constitutionally unable to control his own statements if released from detention:

> Then the war ended – by reason of the factual elimination of the principal Central European belligerents – and most of the fighting came to an end. That statement is, of course, only a convenient conventional approximation. Since there was no Peace Treaty, and no-one with whom a treaty could be made, a state of war was, by declaratory ministerial certification, presumed to be continuing. This approximation to peace was, fittingly enough, signalized by the ceremonial march of four allied commanders-in-chief carrying the incinerated remains of their executed German opposite numbers at the dead of night in an old rubbish bin, to sink it in the waters of the Rhine. An attempt, presumably, to efface ritually the ineffaceable memories of a war that never should have been.[77]

As he bluntly told his son Julian, 'I am painfully aware that I am ashamed of being an Englishman and equally ashamed of anyone who is *not* ashamed of being an Englishman.'[78]

Despite his completion of several manuscripts after the end of the war, no publishers could be found to print them. No doubt this was in part because they all appear to have contained unyielding attacks on Winston Churchill and the British government. He complained that now even *The Times* refused to publish his letters to the editor.[79] On the far right, he was still seen as a kindred spirit and a possible source of funds. Following the creation of the far-right Union Movement by Oswald Mosley and others in 1945, Pitt-Rivers was contacted by leaders seeking money for the cause. If he made a contribution it was probably not substantial, given his own financial difficulties at the time (characteristically, he blamed the Jews and referred in correspondence to the 'Hebrew economy').[80] At the same time, however, he found the funds to host a music festival at his estate in late 1945 in an effort to endear himself once again to the surrounding community. While various local dignitaries attended the festivities and expressed the usual platitudes about its importance, attendance was low and it appears to have had no lasting effect.[81] The fact that the event's host had only recently returned to Dorset from wartime detention, and his pre-war views were still widely known there, must have made the event particularly awkward to attend for some.

Just months after the end of the war, a new woman entered Pitt-Rivers' life. According to one account, the pair met over cocktails at the Ritz in London, with the unusual detail that she was initially the one serving them rather than partaking. The woman was Stella Lonsdale, a champagne-loving brunette with a past that the term 'checkered' could hardly begin to describe. As Matthew Sweet has recently observed, the details of her life are difficult to confirm even today, but the stories she told were literally unbelievable to most she encountered.[82] Born Stella Edith Clive in 1913, she was the daughter of a candy seller who abandoned the family when she was a girl. In 1934, she met a married Danish businessman, Paul Holme, with whom she not only conducted an affair but also managed to convince to bankroll a secretarial bureau for her to run. More men followed, and two years later, she met Nicholas Sidoroff, a highly questionable character with a penchant for gambling, theft and con games. The pair ran off to Monte Carlo, where they married under the false and bombastic identities of 'Prince and Princess Dimitri Magaloff'. After subsequently losing their money at the Monte Carlo casinos, 'Princess Magaloff' sent a cable to Holme asking for help. Foolishly, he agreed to bail out his former lover. The following year, the couple – now living in London and calling themselves the Warners – had a child, Felix, who died just four months later. Sidoroff was soon arrested for illegally entering the country and sent to prison.[83]

The story only got stranger from this point. In prison, Sidoroff met John Lonsdale, a well-known jewel thief, con artist and right-wing gun smuggler who was part of a group known as the 'Mayfair Playboys'. In 1937, Lonsdale had been part of a gang that assaulted a Cartier representative at the Hyde Park Hotel as part of a jewel heist.[84] With her husband and Lonsdale incarcerated together, Stella soon became enamoured with the latter and when he was released from prison she agreed to marry him. Sidoroff offered the obvious objections, but Stella argued that their marriage in Monte Carlo had lacked legal validity. This argument was accepted, the marriage with Lonsdale was permitted and Stella subsequently spent the wedding night in bed with Lonsdale on one side of her and Sidoroff on the other.[85]

With the outbreak of war, John Lonsdale enlisted in the Royal Engineers and was deployed to France. Stella soon followed, arriving in Paris and then heading to Nantes, where her husband was stationed. Her presence was not welcome, in part because she persisted in claiming that she was related to the Minister of War and hinted that she was employed by British intelligence. Hoping to rid himself of this nuisance, Lonsdale's commanding officer sent him back to England, but his wife refused to follow, claiming that Sidoroff, whose loneliness had since led him to contract venereal disease, would be waiting for her if she returned. She refused to leave even when it became clear that Nantes would imminently fall to the Nazis and on 21 June the German army arrived in the city.[86]

At this stage, the facts of Stella's story became essentially unverifiable. MI5 could never conclusively determine exactly why she decided to remain in France when she had the chance to escape. Her own explanation was that she wanted to help the war effort behind enemy lines, and her later interrogators agreed that she was probably enamoured with the idea of being a spy.[87] Officially she took a position as an English instructor at a language school after the occupation began, subsequently meeting a young

Frenchman named Jean Platiau who she claimed was infatuated with her. The details of what transpired between them are unclear, but Stella maintained that Platiau denounced her to the Germans in some kind of effort to secure her affections and to convince her to run away with him. Regardless of the exact reason, the pair was arrested and ended up at a German interrogation facility in Angers.[88]

Once there, she was interrogated by a German officer who she said put out a cigarette on her forehead along with another officer she called 'René' whose true identity she would never reveal. She claimed to have recognized René from her time in Paris years before. He initially showed no sign of recognition, however, and after a brief tribunal hearing she was sentenced to death. Days later he appeared in her cell, confessed that he had recognized her and told her he was disenchanted from the German cause. He then convinced her to save her life by offering her services to German intelligence, after which she could double-cross her handlers and return to England. In addition, he would convince German intelligence of her reliability and give her key pieces of information to pass along in return for anonymity and a British passport. Stella agreed, put on the act he suggested and was subsequently released to become a German agent.[89]

If the forgoing account (again, this was the unverifiable story that Stella herself told) seems implausible, there were more outlandish tales to come. She claimed that René provided her with travel documents so she could move around the country and sent her on various missions to undermine the occupation. After her cover was blown in Paris, René advised her to flee to Marseilles, in ostensibly non-occupied Vichy France, and return to London from there with the help of an underground network. In Marseilles she took several lovers and joined operations against the Germans. MI5 agreed that her associates there 'never had cause to regret the connection' but argued that if she were actually a German agent she might take part in such operations to 'build herself up' to gather more valuable intelligence.[90] She eventually escaped to London through Lisbon, just days after Platiau had been shot in Nantes in an act of collective retribution following the assassination of a German officer.[91]

British intelligence had no idea what to make of this woman when she arrived on their doorstep. Had she been working for the British cause all along, as she claimed, or was she really working for German intelligence? Had she worked for both sides and been turned at some point, and, if so, which side was she working for now? And who was René? There were far more questions than answers. MI5 initially put her up at the Waldorf Hotel and interrogated her at length. Her story was consistent, if implausible, and she offered to work for British intelligence. The interrogators she encountered were beguiled, particularly when she went into lengthy descriptions of her sexual exploits, and they could make little progress on determining the truth of her statements. She received letters from her various lovers but these revealed nothing when they were intercepted and opened by MI5. She managed to convince both Sidoroff and Lonsdale to give her money, but she was no longer romantically interested in either.[92]

In early January, she took an ill-advised trip with friends who worked for the Ministry of Aircraft Production to visit a torpedo development facility, which alarmed MI5 greatly. As a result, she was arrested and detained under Defence Regulation 18B in early 1942 and would remain in custody until the end of the war. By this stage, MI5 had effectively given up on ever discovering her true activities and intentions and simply turned her over to the Home Office to decide whether she was a threat to the war effort. One MI5 agent reported that she was 'not normal' and herself had no idea whether she was telling the truth or not.[93] The 18B Advisory Board examined the evidence in depth and concluded that it too had no idea what she had been doing in France but that she was probably no longer a threat to national security in 1944. She was eventually released in May 1945.[94] Bizarrely, she then took a job at a car repair shop before meeting Pitt-Rivers.

The pair evidently first bonded over their mutual experiences as 18B detainees, though Pitt-Rivers had been released by the time she was detained. Whether or not she ever revealed more about her true activities in France to him, particularly whether she had ever actually been working for the Germans, is not known. Much like Becky Sharpe, Stella initially arrived at Hinton St Mary ostensibly as Pitt-Rivers' secretary but it was obvious that she was always more. Their relationship was tempestuous but enduring, and Stella would remain with Pitt-Rivers – more or less – until the end of his life. Her own libertine sexual attitudes were much in line with his, and among the services she provided was the management of his casual love affairs on the condition that none of them would become permanent and threaten her status in the household. She would never marry Pitt-Rivers but did adopt his surname by deed poll, making her the third and final woman in his life to take his name.

The elder Pitt-Rivers sons were by now embarking on careers of their own. Both had served in the war: Michael in the Welsh Guards, and Julian in the Royal Dragoons, for which he wrote the official regimental history of the conflict.[95] Following the war, Julian became the private tutor to King Faisal II of Iraq, who had only been born in 1935 and was consequently under the tutelage of a regent. Julian's task was to arrange for the boy king's education in England. As a result, in August 1946 the young Faisal visited England and spent several weeks at the Hinton St Mary manor house where he was instructed on the principles of British agriculture and archaeology. Evidently, any concerns about the fact that the estate's owner had only recently been released from detention had been pushed to the side, and a published photograph in the local newspaper showed Pitt-Rivers and Faisal standing next to one another. He subsequently entered preparatory school at Sandroyd School, which was housed in the Rushmore estate house, before going on to Harrow.[96] Julian Pitt-Rivers kept a close watch on the king for the first years of his education, reporting diligently to the Foreign Office about his activities and his views of how the king's education was progressing.[97] Faisal II would later be overthrown and killed in a 1958 coup, marking the end of the monarchy. After a period of instability, the Ba'ath Party launched another coup to take control of the country, and in 1979 a brutal Ba'ath leader named Saddam Hussein became its president.

In 1946, Julian had married Pauline Tennant, the beautiful daughter of aristocrat David Tennant and actress Hermione Baddeley, known for her role in the film *Mary Poppins* among other performances. Like her mother, Pauline was an actress and appeared on stage before moving to Iraq with her new husband. Living in Baghdad high society, the pair once dressed as Oscar Wilde and his lover Lord Alfred Douglas for a costume party, causing heads to turn.[98] The elder Pitt-Rivers was less than impressed with his new daughter-in-law, telling Julian, 'Your little bride is an attractive child and I think really intelligent – provided she does not rate her intelligence too high. She seems quite unaffected at present, but do stop her overdoing the face paint...and hair tinting stuff.'[99] He later described her as 'self-indulgent, untidy and far too undisciplined', though he still claimed to like her.[100]

On the other hand, Pauline appears to have detested her father-in-law. She later remembered him as a tyrant, demanding complete obedience to his household rules. On her second visit to the Hinton St Mary house, she recalled him attempting to gain entry to her bedroom in the night, but she had securely locked the door and left him knocking and demanding entry. The next morning he claimed that he had just wanted to discuss *Weeds in the Garden of Marriage* with her despite the late hour.[101] She found him to be a pretentious windbag, complaining that virtually everything he talked about was 'boring'. At her 1949 birthday party, a combination of strong feelings and alcohol resulted in an open row between them. He acidly wrote to her complaining about her conduct and noting that she had asked for whiskey and brandy during the party. 'It was not new to you that in my house ladies are not offered or allowed whiskey or spirits of any sort – particularly my daughter-in-law', he told her.[102]

Leaving his duties with the king of Iraq, Julian returned with his wife to Oxford to study anthropology. His father was outraged by the fact that he was learning about the field from his former opponents and rivals, including A.R. Radcliffe-Brown, who he accused of stealing his work, and E.E. Evans-Pritchard, the protégé of Malinowski with whom he had clashed in the late 1930s. In 1949, Julian undertook field research in Spain, where he examined rural village life and published his findings in the book *The People of the Sierra*.[103] The introduction was written by Evans-Pritchard and referred to his father's work in *The Clash of Culture* as having 'long been recognised as an original and outstanding study of primitive societies'.[104] Despite this praise, the son had now outdone the father. Julian's marriage with Pauline ended in 1953, and he would soon embark on a sterling academic career that included posts at Berkeley, the University of Chicago and the London School of Economics, among other universities. Ironically, Berkeley and the London School of Economics had been the two universities, along with Oxford, at which his father had once enjoyed his closest connections and had perhaps his best chances of securing a high-flying academic post of his own decades before.

Julian had the flexibility to pursue his career thanks in part to that fact that he was not the eldest son and therefore not saddled with the estate management duties that had helped derail his father's academic work. Instead, it was his elder brother Michael that most of these duties fell upon. Like Shakespeare's King Lear, Pitt-Rivers decided to divide his estate while he was still alive, giving Michael the Rushmore house and the surrounding lands, including the Larmer Tree Grounds. Decades of neglect and the war had reduced these to a state of near dilapidation, forcing Michael to borrow money to undertake renovations.

There was more serious trouble to come, however. In January 1954, Michael was arrested and charged with 'conspiracy to incite certain male persons to commit serious offences with male persons' along with his cousin, Lord Montagu, and Peter Wildeblood, the diplomatic correspondent for *The Daily Mail*.[105] The charges stemmed from a party months earlier at a beach hut belonging to Montagu that included the trio and two RAF airmen. The airmen later claimed that there had been dancing and 'indecent acts' at the hut. When Montagu left for a trip to America, Michael invited them to stay at his estate and to visit him in his London flat, where more 'unnatural acts' were alleged to have taken place.[106] Faced with potential legal sanctions for their involvement, the airmen agreed to testify against their hosts to avoid criminal charges of their own. The arrest and subsequent trial were huge blows to Pitt-Rivers, who evidently had no idea that his son was gay.

The Montagu Case, as it became known, was absurd from the start. The Home Secretary, Sir David Maxwell Fyfe, had sworn to 'rid England of this plague' and hundreds of men had already been sent to prison on suspicion of homosexuality.[107] In their reports to the police, the RAF officers accused two dozen other men of similar offences but these were the only three prosecutions that resulted. The details and circumstances certainly seemed to fit the description of a witch-hunt.[108] Furthermore, even if the charges had been true, everyone involved in the case had been a consenting adult and there was never an accusation of assault. In essence, the three men were being put on trial for victimless private conduct. The trio denied the charges but were convicted in March, with Michael Pitt-Rivers and Wildeblood, the latter of whom had openly admitted in court that he was gay, receiving eighteen months in prison each. Montagu received a lesser sentence.[109] The men had lost the case but won in the court of public opinion by generating sympathy for their plight. Following the announcement of the verdict, a crowd outside the court cheered the three as they were taken away. Wildeblood subsequently wrote a book about his experiences and in 1957 a committee led by Lord Wolfenden endorsed the de-criminalization of homosexuality between consenting adults.[110] The recommendation became law ten years later.

The Montagu Case had taken a toll on Pitt-Rivers. He now tried to convince Michael to give up the Rushmore estate, making offers to purchase him land elsewhere. When these were refused, their relationship suffered a major breach. In 1958, Michael married Sonia Orwell, the widow of writer George Orwell and the executor of his literary estate. Predictably, the marriage was tempestuous and did not last.[111] Michael then committed himself to estate management and writing, making improvements and renovations to the physical legacy of General Pitt-Rivers. In 1966 he wrote *Dorset. A Shell Guide*, which was successful and regarded as one of the better editions in the travel guide series.[112]

George Pitt-Rivers was now in the last years of his life. His health was deteriorating gradually, and both his asthma and his war wound were becoming more troublesome. The Montagu Case and the subsequent feuds surrounding it had poisoned his relationship with his two eldest sons, as Julian had supported his brother throughout the ordeal. Financial pressure led to parts of the estate being sold off. In the early 1950s, Pitt-Rivers attempted to retrieve some of the objects he had sent to the Pitt-Rivers Museum in Oxford during his fieldwork in the early 1920s. The items had mostly been in storage since arriving there, but the university now refused to return them on the grounds that he had legally lost possession by leaving them there for so long. The ensuing threats and legal wrangling went on for years.

The Pitt-Rivers Museum in Farnham was even more troublesome, however. Still the private possession of the family, it was losing money and required costly repairs that Pitt-Rivers could not readily afford. In the early 1960s, he arranged for the collection to be severed from his main estate and transferred to a separate trust run by Stella. He made it clear that the trustees were expected to negotiate for the donation of the collections to the nation after his death. By 1965, however, objects from the museum had begun to appear at art and antiquities auctions in the United States without a listed provenance.[113] The collection was rapidly being sold off at Stella's behest.

In these troubled contexts, Pitt-Rivers made his final intervention in the academic realm. In 1963, a blacksmith down the road from his house in Hinton St Mary uncovered a spectacular Roman mosaic. It was subsequently determined that it had probably been part of a large villa complex and was lifted for preservation. In the centre of the mosaic was the image of a man with the Greek letters Chi and Rho overlapping behind his head. These letters, coupled with imagery elsewhere on the mosaic, led some scholars to identify it as an image of Christ, which would make it one of the earliest known representations of Jesus in the entire Roman world.[114] Others disagreed, identifying it as the Emperor Constantine.

Though already unwell, Pitt-Rivers felt obligated to intervene in the debate, particularly because the mosaic had been found on land he formerly owned. He embarked on a study of Christian imagery, publishing his findings in a short and heavily illustrated book entitled *The Riddle of the 'Labarum' and the Origin of Christian Symbols*.[115] It was less an academic study than an attack on religion as a whole with a few choice passages attacking the Jews as well. It received little serious attention, and one reviewer, Oxford theologian Henry Chadwick, bemoaned that it was 'argued very inadequately'. 'Mr. Pitt-Rivers was a veteran anthropologist. This excursion into historical research will be deeply regretted by his admirers.... It is painful to see a respected scholar burning his fingers', Chadwick continued.[116] If anything, this description of Pitt-Rivers as a 'respected' scholar was probably too generous at this point.

By this time, Stella herself had developed a new romantic interest. In 1963, while she was still Pitt-Rivers' companion, she married Raoul Maumen, a Frenchman with a questionable past that she claimed to have known from the French Resistance but in reality had been a racketeer during the war.[117] She completely concealed the marriage from Pitt-Rivers, justifying her long visits to France on the grounds that she was receiving unspecified medical treatment. The Maumens eventually moved into a flat in London together. Pitt-Rivers soon found himself being convinced by Stella to take long voyages around the world to tropical climes, ostensibly for his health but in reality to get him out of England.

Despite already being married to another man, Stella was with Pitt-Rivers when he died on 17 June 1966. His obituary in *The Times* memorialized him as the 'owner-director of the Pitt Rivers Museum at Farnham', recounted his anthropological work in the 1920s and incorrectly said that he had been married three times. There was no mention at all of his wartime internment.[118] For decades to come, a small memorial bearing his name appeared every year in the paper around the anniversary of his death. It read simply: 'Constantly in the thoughts of his friends and Stella.'[119] Years later, she would establish a more lasting memorial to Pitt-Rivers in a place where it could hardly be overlooked.

AFTERWORD: A LIFE OF CONTRADICTIONS

As had been the case when General Pitt-Rivers died in 1900, George Pitt-Rivers' heirs were faced with difficult decisions following his death. The Farnham museum was no longer open to the public and in desperate shape. Objects from its collections had already begun to disappear into private hands. In 1972, the sales were disclosed in *The Times*, and in 1974, Stella announced that the remainder of General Pitt-Rivers' collection would be given to the nation and housed at Salisbury Museum. After extended negotiations, the Treasury eventually accepted this offer and the objects were moved to the museum, where they remain to the present day.[1] The Pitt-Rivers Museum in Farnham thus ceased to exist and was subsequently converted into private residences.

Michael Pitt-Rivers opposed the dispersal of the collections but in the end decided that nothing could legally be done to stop it. For the remainder of his life he managed the Rushmore estate, making substantial improvements to the Larmer Tree Grounds and the surrounding area. In addition to the Dorset *Shell Guide*, he wrote a number of book reviews and shorter works but, regrettably, appears to have never embarked on a substantial autobiographical project.[2] As his father's eldest son and presumptive heir, he had been dragged across Europe on his quixotic travels with Becky Sharpe in the pre-war years and had been forced to spend time with Alfred Ploetz and his circle in Munich. His memories and reflections of those years and experiences would have been a fascinating and valuable source for historians. In 1995, he reopened his great-grandfather's restored Larmer Tree Grounds to the public following extensive renovations. He died in 1999.[3]

As already noted, Pitt-Rivers' second son, Julian, went on to enter the same academic field as his father and to far surpass him in achievements. Following academic posts at research universities in the United States, he taught at the London School of Economics until 1977 and then moved permanently to France. By the 1980s he was recognized as one of the most significant scholars of Iberian kinship, mythology and the tradition of bullfighting. Having completely disavowed his father's racist views, he was seen as one of the founders of social anthropology in Spain and was admitted to the prestigious Order of Isabella the Catholic. He died in 2001.[4]

One of the most remarkable aspects of George Pitt-Rivers' life was the array of talented women that passed through it at various stages. His first wife, Rachel, pursued an acting career after their marriage ended and appeared in dozens of television programmes and films under her stage name until the mid-1960s. She remained a strong supporter of her sons, campaigning for the de-criminalization of homosexuality following Michael's conviction. She died in 1979. His second wife, Rosalind, went on to a sterling and high-flying academic career after their separation, earning a PhD in 1939 and embarking on studies of the thyroid. At the end of the war, she was sent to Belsen Concentration Camp to conduct a nutritional study on starving and dying inmates there. The horrors she witnessed remained with her for the rest of her life.[5] She soon became a world-renowned thyroid hormone researcher and contributed to multiple books on the subject. In 1954, she was elected to the Royal Society and in 1979 served as president of the European Thyroid Association. After retiring to Dorset, she died in 1990.[6] Her son, Anthony Pitt-Rivers, eventually moved into the Hinton St Mary manor house and, at the time of writing, is the last living son of George Pitt-Rivers.

The third and final woman to take Pitt-Rivers' name was, of course, the only one who did not actually marry him. By the time of his death, Stella Pitt-Rivers had acquired the surname of Maumen thanks to her marriage to a Frenchman with a scurrilous reputation and questionable connections. Following his death, the couple left Britain and moved to a vineyard in Provence. In the early 1970s, however, Raoul Maumen turned up at Michael Pitt-Rivers' door asking for help and bearing a piece of family silver to establish the veracity of his connection to Stella. He claimed that Stella was threatening him and asked for whatever protection could be offered. When he was told that nothing could be done, he returned to France and was found dead in his bathtub soon after. It turned out that the bottle of whiskey he had been drinking that evening contained a strong dose of household disinfectant. Whether or not Maumen took his own life was never fully established, but there were rumours that he might have been assisted in the process by his less-than-savoury acquaintances.[7] Either way, no one ever faced criminal charges in his death.

In 1990, Stella created a lasting memorial for the man she refused to marry by endowing an academic chair and laboratory at Cambridge University. The George Pitt-Rivers Professorship of Archaeological Science was thus created, as was the George Pitt-Rivers Laboratory for Bioarchaeology. According to its website, research projects since the lab's creation have examined 'palaeolithic plant foods, tree exploitation in arid zones, Greco-Roman agriculture, and Eurasian crop plant movement'.[8] A bust of Pitt-Rivers himself stands in the lab – perhaps a fitting reminder of the fact that his greatest scientific legacy has undoubtedly come from the endowed facility and professorship that bear his name rather than own work. Stella herself subsequently returned to France and descended into alcoholism. She returned to England at the end of her life and died in 1993. She apparently never revealed the name of the mysterious René who had supposedly saved her life in France, if he ever existed at all.

Yet, perhaps the most unusual and remarkable story belongs to Catherine 'Becky' Sharpe, Pitt-Rivers' erstwhile mistress in the 1930s. Arriving in South Africa as war broke out, she married Lance Taylor, an Oxford-educated doctor who was the youngest son of Randlord J.B. 'Lucky Jim' Taylor, formerly one of the richest men in the country. After a brief honeymoon, the couple returned to England to allow Lance to assist the war effort as a physician. His wife soon gave birth to a baby boy, and with the German invasion scare approaching, he insisted that his wife and child return to South Africa for their safety. Disliking the gaudy parties and decadence of the Taylor family estate while her husband was facing German bombs in London, Catherine insisted on returning to England and boarded a neutral Portuguese ship in April 1941. Off the coast of Freetown, her infant son fell ill and died. He was buried on the Portuguese island of Madeira.[9]

Devastated by the loss, she pressed on to Lisbon, where she hoped to catch a plane to Bristol. Even with a letter from Jan Smuts in her pocket – he was a long-time friend of the Taylor family – there was nothing to do but wait for weeks until a seat became available. Finally, she managed to get aboard a flight only to be greeted by MI5 agents at the airport. The Secret Intelligence Service (MI6) had sent information to London indicating that 'a woman with good legs, not a Portuguese' would shortly be arriving on a flight from Lisbon. The woman would be a German agent, and she would be carrying bank notes with instructions from her handlers written into the margins. MI5 reported that three women on the flight fit the description but Taylor was 'the most suspect'. Under questioning, she admitted to having been Pitt-Rivers' secretary before the war but 'nothing could be found on her of an incriminating nature', so she could not be detained. 'She is the subject of further investigation', MI5 counterespionage chief Guy Liddell recorded in his diary.[10]

Catherine Taylor soon had two more sons, and at the end of the war the family returned to South Africa. In 1948, she entered politics as part of the United Party, the party of Jan Smuts that had been defeated by the Afrikaner-dominated National Party in the general election that year. She was strongly opposed to the apartheid system and the suppression of the African vote that the National Party soon embarked upon. She was particularly known for her opposition to the absurdity of the racial classification system and for helping constituents who had suffered through its vagaries. In 1953, she became one of the very few women elected to parliament and on 6 September 1966 she coincidentally rode in a lift with the knifeman who would assassinate Nationalist Prime Minister Hendrik Verwoerd just moments later.[11] After the 1970 general election, she became Shadow Minister of Education, which gave her a portfolio that included student affairs at a time when South African universities, like many around the world, were in turmoil.

Disgusted by the political infighting of the United Party and its inability to make substantial inroads against the Nationalists and the apartheid system, Taylor resigned from parliament in 1974. She travelled extensively and penned an autobiography, which was edited and published as a political rather than personal memoir in 1976 and included only brief mentions of her time as Pitt-Rivers' secretary. There was no suggestion that their relationship had been anything more than purely professional, and she refrained from extensively describing her experiences with him in Nazi Germany beyond a brief mention of meeting Hitler and a description of her time in the Hitler Youth camp.[12] The AGF did not appear in her memoir at all.

For a woman who had been immersed in the heated climate of the British far right for years, Taylor's views and actions after the war seem incongruent. It is possible that in her early twenties she simply never considered the wider implications of the things she was hearing from Pitt-Rivers and in the meeting rooms of the AGF. It is perhaps telling that in no surviving correspondence does she appear to have personally expressed the pro-Nazi or anti-Semitic views of her employer. Even her *Anglo-German Review* article about the Hitler Youth camp she had visited focused on the national pride and discipline on display there and had none of the anti-communist or anti-Semitic rhetoric common to Pitt-Rivers' writings. It is also possible that her early and immature views were changed by the destruction of the war and the suffering that resulted from Nazism. Either way, the consummate politician Catherine Taylor had little resemblance to the young Becky Sharpe that had accompanied Pitt-Rivers across Europe in the 1930s and there is no evidence that the two of them ever met again after her departure in 1939. She died in 1992, just two years before the first free elections in South Africa ended the apartheid system she despised.

In the end, George Pitt-Rivers had been professionally outdone by virtually everyone around him. Both of his wives were accomplished women, particularly given the relatively few opportunities available to women at the time, and his son Julian enjoyed the academic career that should have been his. His former mistress had moved on to a high-flying political career and spent time on the international lecture circuit. Indeed, by the early 1950s, most of the intellectual circles in which Pitt-Rivers had enjoyed influence were gone. C.G. Seligman died in 1940; Bronislaw Malinowski died in the United States in 1942; Oscar Levy died in 1946; and Arthur Keith died at the age of 88 in 1955. Despite years of silence during their political separation in the later 1930s, Levy and Pitt-Rivers had corresponded throughout the war until nearly the end of the former's life. The generation that dominated anthropology after 1945 were the figures that Pitt-Rivers detested – Firth, Evans-Pritchard and young scholars like Margaret Mead, who had been influenced by Franz Boas, one of the targets of his ire in the Race and Culture Committee report – and as a result he had no home in anthropology after 1945. His unyielding anti-Semitism alone would have precluded him from serious academic activity after the war.

One of the few to openly maintain similar views in the post-war world was Pitt-Rivers' friend Reginald Ruggles Gates, who took a position at a historically African American university in the United States so he could scientifically study the students. The inevitable controversy over his views and intentions led to his dismissal from the university, for which he blamed the Jews.[13] He corresponded with Keith until the latter's death, and in one of Keith's final letters, he confessed that he no longer even recognized the path that British anthropology had taken and bemoaned that his concept of race had been dispensed with entirely by post-war anthropologists. Both men seem to have viewed the Jews as responsible for this until the end of their lives.[14] Pitt-Rivers surely agreed with their assessment.

In the final reckoning, the book George Pitt-Rivers always should have embarked upon was a life of his grandfather, General Pitt-Rivers. While he proved willing to assist the BBC with a 1953 radio programme about his famous forebear, he never put pen to paper on the topic. Ultimately, this was the one book that only he himself could write: his son Michael later recalled that he had jealously locked away the general's personal papers and other materials in the Farnham museum for such a project – 'and of course never wrote a word'.[15] But why?

The answer lies in George Pitt-Rivers' character. He was fundamentally a man driven, and perhaps dominated, by his ego. He was obsessed with a bygone age – the world of the general, of which he had tasted hints as a child and as a young man on leave from the Royal Dragoons – and all of his writings after the First World War were focused on the idea that the world was getting worse and heading towards cataclysm. He characteristically blamed the Jews for much of the danger he diagnosed, claiming that they were responsible for the rise of Bolshevism and virtually every other malady, as he saw it, of the modern world. In his view, the events of 1914 had driven a stake into the heart of whatever remained of the old world, and without radical countervailing action, things would only continue to get worse.

The solution to the world's ills, he believed, lay first in the philosophy of Friedrich Nietzsche and then with Adolf Hitler and National Socialism. His enthusiasm for the Nazis went far beyond the pale and took him into circles of people far more fanatical and dangerous than himself. There is no evidence that he ever intended to become a traitor or sell out his country, but it is inarguable that he unwittingly helped the German cause in the 1930s through his publications and public statements. His internment under Defence Regulation 18B was explicitly predicated on the idea – probably not an incorrect one, incidentally – that he would be unable to control his tongue if allowed full freedom in wartime.

Fundamentally, Pitt-Rivers believed that with the defeat of Nazism in 1945, there was no longer hope for the world. His few post-war writings were even more pessimistic than his earlier works and his private letters indicate a clear despair about the future of the world. This, of course, was predicated on his anti-Semitism and his intrinsically racist and classist views. His own familial and financial problems after 1945 no doubt compounded his despondency, and the rupture with his eldest sons after the Montagu Case took a toll as well. One gets the impression that after the mid-1950s he had essentially given up on everything. He had hoped in the end to preserve his grandfather's museum in Farnham, but even that was taken from him without his knowledge.

In the final reckoning, Pitt-Rivers had squandered incredible potential. He could have easily enjoyed a career similar to, or even exceeding, that of his son. Instead he became consumed by cultural pessimism, and when Hitler seemed to offer a political solution to the ills he claimed to have discovered, he embraced Nazism with a fanaticism that blinded him to everything else. His interest in eugenics, a field in which he gained fairly significant international recognition, was predicated on the ideas of racial difference that he had been holding unmodified since 1920 even after legitimate science itself was increasingly calling them into question. He never really moved beyond the politics of the 1930s, and after the war he was both marginalized and intellectually incapable of engaging with the new direction the world had taken. He never undertook the biography of his grandfather because he believed the world had thrown away its chances to salvage what was left of that era and descended instead into an age that he could hardly begin to understand. In the end, his lack of action was the last protest he could muster against a world that he believed had rejected not only him but also everything for which he stood.

NOTES

Introduction

1 'Their finest hour' speech, Winston Churchill, 18 June 1940 (Chartwell Papers 9/140A/32-55).

2 8 July 1940 memorandum from the Dorset Constabulary to MI5 (Security Service files KV 2/831).

3 Barry Domvile Diaries, National Maritime Museum, London, v. 56, 27 June 1940 entry.

4 8 July 1940 memorandum from the Dorset Constabulary to MI5 (KV 2/831).

5 Malcolm Muggeridge, *Chronicles of Wasted Time: The Infernal Grove*, vol. 2 (London: Fontana, 1975), 126.

6 A.W. Brian Simpson, *In the Highest Degree Odious: Detention without Trial in Wartime Britain* (Oxford: Clarendon Press, 1992), 217–218.

7 See, for instance, Mark Bowden, *Pitt Rivers: The Life and Archaeological Work of Lieutenant-General Augustus Henry Lane Fox Pitt Rivers, DCL, FRS, FSA* (Cambridge; New York: Cambridge University Press, 1991).

8 Specific descriptions found in letter from Winifred to Pitt-Rivers, 6 September 1940 (George Pitt-Rivers Papers, Churchill Archives Centre, Cambridge 18/1).

9 Arthur Keith, 'Introduction', in *Weeds in the Garden of Marriage*, ed. George Henry Lane Fox Pitt-Rivers (London: N. Douglas, 1931), x.

10 Kazuo Ishiguro, *The Remains of the Day*, 1st American ed. (New York: Knopf: Distributed by Random House, 1989).

11 George Henry Lane Fox Pitt-Rivers, *The Riddle of the 'Labarum' and the Origin of Christian Symbols* (London: Allen & Unwin, 1966).

12 Patrick Wright, *The Village That Died for England: The Strange Story of Tyneham* (London: Vintage, 1996), 162–164.

13 MI5 reports concerning Pitt-Rivers, 1940 (KV 2/831).

14 MI5 reports concerning Pitt-Rivers and property recovered in his home, 1940 (KV 2/831).

15 Wright, 163.

16 Letter to Pitt-Rivers from Ulric Thynne, 19 March 1942 (George Pitt-Rivers Papers, Churchill Archives Centre, Cambridge 23/3).

17 As of 8 June 2014, entries exist for General Pitt-Rivers, Baron Forster, Catherine D. Taylor, Rosalind Pitt-Rivers, Michael Pitt-Rivers and Julian Pitt-Rivers.

18 George Pitt-Rivers, *The Story of the Ancient Manor of Hinton St Mary* (Farnham: The Pitt-Rivers Museum, 1947).

Chapter One

1   An account of Pitt-Rivers and Petrie meeting can be found in R. Burleigh
    and J. Clutton-Brock, 'Pitt Rivers and Petrie in Egypt', *Antiquity* 56
    (1982): 208–209; Alice Stevenson, '"We Seem to Be Working in the Same
    Line": A.H.L.F. Pitt-Rivers and W.M.F. Petrie', *Bulletin of the History of
    Archaeology* 22, no. 1 (2012): 4–13; Mark Bowden, *Pitt Rivers: The Life
    and Archaeological Work of Lieutenant-General Augustus Henry Lane Fox
    Pitt Rivers, DCL, FRS, FSA* (Cambridge; New York: Cambridge University
    Press, 1991), 92–93.
2   Bowden, 3–4.
3   Ibid., 4.
4   Ibid., 6–7.
5   Ibid., 15.
6   Ibid., 14–15.
7   Ibid., 15.
8   Ibid., 17–18.
9   Ibid., 19–22.
10 Ibid., 23.
11 Ibid., 23–24.
12 Michael Pitt-Rivers, 'Cultural General', *Books and Bookmen* 22,
    no. 9 (1977): 24–25.
13 Bowden, 31–33.
14 Pitt-Rivers, 'Cultural General', 25.
15 Bowden, 29–31.
16 Ibid., 49–50.
17 Ibid., 48–77.
18 Ibid., 90–92.
19 Ibid., 94.
20 Ibid., xiii, 103.
21 Ibid., 37.
22 Ibid., 31.
23 George Pitt-Rivers, *The Story of the Ancient Manor of Hinton St Mary*
    (Farnham: The Pitt-Rivers Museum, 1947), 26–27.
24 Bowden, 103.
25 Ibid., 110–120.
26 Ibid., 125–126.
27 Ibid., 33.
28 Ibid., 35.
29 Ibid., 36. See also Desmond Hawkins, *Concerning Agnes: Thomas Hardy's
    'Good Little Pupil'* (Gloucester: A. Sutton, 1982).
30 Bowden, 52–53.
31 Ibid., 52.
32 Ibid., 52–53.
33 Ibid., 98–101.
34 Ibid., 142–143.
35 Ibid., 145–147.
36 Pitt-Rivers, *The Story of the Ancient Manor of Hinton St Mary*, 24–25.

37 Bowden, 149.
38 Ibid., 150.
39 Ibid., 154.
40 Ibid., 155.
41 Ibid., 160–165.
42 Debbie Challis. *The Archaeology of Race: The Eugenic Ideas of Flinders Petrie and Francis Galton* (London; New York: Bloomsbury Academic, 2013).
43 Ibid., 36–37, 152.
44 Pitt-Rivers, *The Story of the Ancient Manor of Hinton St Mary,* 26; Bowden, 39.
45 Pitt-Rivers, *The Story of the Ancient Manor of Hinton St Mary,* 26.
46 Ibid.
47 Letter from Pitt-Rivers to Malinowski, 4 September 1927 (Malinowski Papers, MS 19, Yale University Library Manuscripts and Archives, Series I, Box 6, Folder 504).
48 Pitt-Rivers, 'Cultural General', 23.
49 Almeric William FitzRoy, *Memoirs*, 6th ed., 2 vols., vol. 1 (London: Hutchinson, 1925), 326.
50 See Richard Carr and Bradley W. Hart, 'Old Etonians, Great War Demographics and the Interpretations of British Eugenics, C.1914–1939', *First World War Studies* 3, no. 2 (2012): 217–239.
51 'How far socialistic ideals are practicable?' essay (Pitt-Rivers Private Papers).
52 'Record of services of 2nd Lt. George Henry Lane-Fox Pitt-Rivers', pp. 44–45 (WO 76/3).
53 'Coming of Age Festivities: Mr. G.H. Pitt-Rivers of Rushmore', *The Western Gazette,* 14 July 1911 (George Pitt-Rivers Papers, Churchill Archives Centre, Cambridge (hereafter GPR) 23/3).
54 Ibid.
55 Ibid.
56 Ibid. See also articles from *The Three Shires Advertiser*, 15 July 1911 (GPR 23/3).
57 'Coming of Age Festivities: Mr. G.H. Pitt-Rivers of Rushmore', *The Western Gazette*, 14 July 1911 (GPR 23/3).
58 'Record of services of 2nd Lt. George Henry Lane-Fox Pitt-Rivers', pp. 44–45 (WO 76/3).
59 Wessel Visser, 'The South African Labour Movement's Responses to Declarations of Martial Law, 1913–1922', *Scientia Militaria: South African Journal of Military Studies* 31, no. 2 (2003): 143.
60 Ibid.
61 Ibid., 143–144.
62 'Serious Riots in Johannesburg', *Wanganui Chronicle*, 7 July 1913, 5.
63 Ibid.
64 C.T. Atkinson, *History of the Royal Dragoons, 1661–1934* (Glasgow: Printed for the Regiment by R. Maclehose and Co Ltd. at the University Press, 1934), 387–388.
65 Visser, 145–146.
66 Atkinson, 389.
67 'Serious Riots in Johannesburg'.
68 See GPR 22/2.
69 Visser, 146–149.

70 There is a vast historiography on the First World War that is far too extensive to recount here. For several recent perspectives, see Eric Dorn Brose, *A History of the Great War: World War One and the International Crisis of the Early Twentieth Century* (New York: Oxford University Press, 2010), 35–36; Christopher M. Clark, *The Sleepwalkers: How Europe Went to War in 1914*, First US ed. (New York: Harper, 2013).
71 Brose, 36; Clark.
72 Brose, 4.
73 Adrian Gregory, *The Last Great War: British Society and the First World War* (Cambridge; New York: Cambridge University Press, 2008), 31.
74 Ibid., 26–28.
75 C.P. Blacker, *Have You Forgotten Yet? The First World War Memoirs of C.P. Blacker*, ed. J.G.C. Blacker (London: Leo Cooper, 2000), 33.
76 Atkinson, 391.
77 Ibid., 391–392.
78 Blacker, 56–61.
79 Brose, 66–67.
80 Ibid., 67.
81 Atkinson, 399.
82 Ibid., 400–403.
83 Ibid., 402–403.
84 Ibid., 409.
85 Ibid., 412.
86 Ibid., 413.
87 Ibid., 415.
88 Ibid., 414.
89 See letter from Oscar Levy to Pitt-Rivers, 1 September 1941 (GPR 18/2).
90 Almeric William FitzRoy, *Memoirs*, 6th ed., vol. 2 (London: Hutchinson, 1925), 531.

Chapter Two

1   For further analysis, see Richard Carr, 'The Phoenix Generation at Westminster: Great War Veterans Turned Tory MPs, Democratic Political Culture, and the Path of British Conservatism from the Armistice to the Welfare State' (Unpublished PhD thesis, University of East Anglia, 2010).
2   Letter by Pitt-Rivers sent from B.E.F., 17 August 1918 (GPR 19/3).
3   Autobiographical notes by GPR for *Who's Who*, early 1930s (GPR 16/6).
4   C.T. Atkinson, *History of the Royal Dragoons, 1661–1934* (Glasgow: Printed for the Regiment by R. Maclehose and Co Ltd. at the University Press, 1934), 414.
5   Ibid., 427.
6   Autobiographical notes by GPR for *Who's Who*, early 1930s (GPR 16/6).
7   George Pitt-Rivers, *The Story of the Ancient Manor of Hinton St Mary* (Farnham: The Pitt-Rivers Museum, 1947), 28.
8   Atkinson, 468–469.
9   Letter from Pitt-Rivers to Commanding Officer, 6th Reserve, 10 March 1919 (GPR 18/1).

10 Atkinson, 470.

11 Letter from Pitt-Rivers to Leopold Maxse, 1 February 1918 (GPR 19/3).

12 See David S. Thatcher, *Nietzsche in England, 1890-1914: The Growth of a Reputation* (Toronto: University of Toronto Press, 1970).

13 H.L. Mencken, *The Philosophy of Friedrich Nietzsche* (London: T.F. Unwin, 1908).

14 Philip Mairet, *A. R. Orage: A Memoir* (London: J.M. Dent and Sons, Ltd, 1936), 21.

15 Ibid., 30.

16 Ibid., 31.

17 Ibid., 47.

18 One of the few to consider Levy's influence is Dan Stone, 'An "Entirely Tactless Nietzschean Jew": Oscar Levy's Critique of Western Civilization', *Journal of Contemporary History* 36, no. 2 (2001). See also Jennifer Ratner-Rosenhagen, *American Nietzsche: A History of an Icon and His Ideas* (Chicago, IL; London: University of Chicago Press, 2012). Much of Levy's work and many of his personal correspondence are gradually being published in German at the time of writing: see, for instance, Oscar Levy, Leila Kais, Steffen Dietzsch and Julia V. Rosenthal, *Gesammelte Schriften und Briefe*, 1. Aufl. ed. (Berlin: Parerga, 2005).

19 Letter from Henry Bergen, Carnegie Institution of Washington, to the Home Secretary, 27 October 1920 (HO 382/93/2).

20 Quoted in Stone, 272.

21 Oscar Levy, *The Revival of Aristocracy*, trans., Leonard A. Magnus (London: Probsthain & Co., 1906), 119.

22 Oscar Levy, 'The Nietzsche Movement in England: A Retrospect, a Confession, and a Prospect', *The New Age* XII, no. 7 (1912): 157.

23 This account of Levy's views is thanks to Stone's path-breaking work in Stone, 276–277.

24 Oscar Levy, 'Nietzsche and the Jews II', *The New Age* XVI, no. 8 (1914): 195.

25 9 March 1921 letter from Levy to Pitt-Rivers (GPR 19/2).

26 For example, Levy, *The Revival of Aristocracy*.

27 Email from Mrs Emma Goodrum, Archivist, Worcester College, Oxford, to the author, 11 June 2014.

28 George Pitt-Rivers, *Conscience & Fanaticism: An Essay on Moral Values* (London: W. Heinemann, 1919).

29 Letter from W.M. Heinemann to Rachel Pitt-Rivers, 15 August 1918 (GPR 21/2).

30 Review in *The Times Literary Supplement*, 3 April 1919 (GPR 30/3).

31 Pitt-Rivers, *Conscience and Fanaticism*, 101–101.

32 Ibid., 108.

33 Ibid., 107.

34 Reviews found in GPR 30/3.

35 Sigmund Freud and A.A. Brill, *Totem and Taboo: Resemblances between the Psychic Lives of Savages and Neurotics* (London: G. Routledge & Sons, 1919); Sigmund Freud and Joan Riviere, *Civilization and Its Discontents*, The International Psycho-Analytical Library (London: Hogarth Press, Institute of Psycho-Analysis, 1930).

36 Pitt-Rivers, *Conscience and Fanaticism*, 44–45. Paragraphing dissolved.

37 Anthony M. Ludovici, 'On "Conscience and Fanaticism"', *The New Age* 25, no. 24 (1919).

38 George Pitt-Rivers, 'The Sick Values of a Sick Age: A Response to Anthony Ludovici', *The New Age* 26, no. 2 (1919).

39 Letter from Levy to Pitt-Rivers, 16 May 1920 (GPR 19/2).

40 Letter from Levy to Pitt-Rivers, 9 June 1920 (GPR 19/2).

41 George Pitt-Rivers, *The World Significance of the Russian Revolution* (Oxford: Basil Blackwell, 1920).

42 Letter from Levy to Pitt-Rivers, 22 June 1920 (GPR 19/2).

43 Stone, 281–282.

44 Oscar Levy, 'Prefatory Letter', in *The World Significance of the Russian Revolution*, ed. George Pitt-Rivers (Oxford: Basil Blackwell, 1920), ix.

45 Ibid., x–xi.

46 Ibid., xii–xiii.

47 Pitt-Rivers, *The World Significance of the Russian Revolution*, 14–15, 44.

48 Ibid., 10.

49 Ibid., 20. Emphasis original.

50 Ibid., 39.

51 Ibid., 39–40.

52 Ibid., 24–25.

53 Ibid., 27.

54 Ibid., 44.

55 22 September 1920: Levy to GPR (GPR 19/2).

56 Ibid.

57 Letter from Levy to Pitt-Rivers, 13 October 1920 (GPR 19/1). Reviews found in GPR 30/3.

58 F.T. Dalton, 'New Books & Reprints: Political', *The Times Literary Supplement*, no. 976 (1920): 638.

59 Letter from Levy to Pitt-Rivers, 27 October 1920 (GPR 19/2). Letter by Clement Salaman in *The Spectator*, 16 October 1920 (GPR 30/3).

60 Clipping from *Birmingham Gazette*, 9 September 1920 (GPR 30/3).

61 Review quoted in promotional materials by Basil Blackwell, 1920 (GPR 30/3).

62 The report of this incident made Levy noticeably uncomfortable: 'I confess that I could have dispensed with these tears. I like women to cry about their husbands, lovers and children – but not about prefaces, pamphlets and poetries', he told Pitt-Rivers (22 September 1920: Levy to Pitt-Rivers, GPR 19/2, 27 October 1920 letter from Levy to Pitt-Rivers, GPR 19/2).

63 Letter from Levy to Pitt-Rivers, 13 October 1920 (GPR 19/2).

64 *The Dearborn Independent,* 1920 (GPR 30/3).

65 J.H. Clarke, 'Democracy or Shylocracy', *Plain English*, 21 January 1922 (GPR 30/3). On the *Protocols* themselves, see Norman Cohn, *Warrant for Genocide: The Myth of the Jewish World-Conspiracy and the Protocols of the Elders of Zion* (London: Eyre & Spottiswoode, 1967); Sergei Aleksandrovich Nilus and Victor Emile Marsden, *Protocols of the Meetings of the Learned Elders of Zion* (London: The Britons, 1931).

66 Stone, 284.

67 Almeric William FitzRoy, *Memoirs*, 6th ed., vol. 2 (London: Hutchinson, 1925), 735.

68 Letter from Levy to Pitt-Rivers, 22 June 1920 (GPR 19/2).

69 6 October 1921 Home Office minutes relating to Dr. Oscar Levy case, HO 382/93/3.

70 Extract from *The Hidden Hand*, October 1921 (HO 382/93/3).
71 Letter from Pitt-Rivers to W. Haldane Porter, 15 February 1921 HO 382/93/3).
72 Letter from Pitt-Rivers to Cecil Harmsworth, 13 July 1920 (GPR 19/2).
73 Letters contained in HO 382/93/1-5, petition from Bloomsbury residents in HO 382/93/3.
74 Letter from Levy to Pitt-Rivers, 9 March 1921 (GPR 19/2).
75 Letter from Home Office to Lord Stamfordham in response to query by the King, 4 October 1921 (HO 382/93/3).
76 Letter from Director of Intelligence to Home Office, 16 April 1920 (HO 382/93/5).
77 Letter to Sir Basil Thompson, Director of Intelligence, 17 June 1920 (HO 382/93/5).
78 Letter from Special Branch to the Home Office, 7 March 1920 (HO 382/93/5).
79 Excerpt from *Jewish World*, 21 November 1921 (HO 382/93/3).
80 Letter from Levy to Pitt-Rivers, 8 February 1922 (GPR 19/2).
81 Letter from Pitt-Rivers to Lady Maudsley, 23 May 1922 (GPR 21/2).

Chapter Three

1 Letter from Pitt-Rivers to Levy, 15 January 1921 (GPR 19/2).
2 George Pitt-Rivers, 'A Psychological Study of the Artist and His Art', in *The Art of George W. Lambert, A.R.A* (Sydney: Art in Australia limited, 1924), 31–40.
3 Chris Cunneen, 'Forster, Sir Henry William (1866–1936)', National Centre of Biography, Australian National University, http://adb.anu.edu.au/biography/forster-sir-henry-william-6213/text10681 (accessed 3 June 2014).
4 C.M.H. Clark, *A History of Australia: Vol. 6* (Melbourne: Melbourne University Press, 1987), 160–161.
5 Letter from Basil Blackwell to Pitt-Rivers, 25 January 1921 (GPR 21/2).
6 Letter from Ian D. Colvin to Pitt-Rivers, 14 November 1920 (GPR 21/1).
7 Offprint of letter to *The Morning Post,* 10 January 1920 (GPR 21/5).
8 Letter from C.G. Jung to Pitt-Rivers, August 1921 (GPR 18/3).
9 'Memorandum on the revolutionary movement in Australia' by Pitt-Rivers, April 1921 (GPR 21/1).
10 Ibid., pp. 1–2.
11 Ibid., pp. 37–41. Pitt-Rivers recommended his friend Elton Mayo for the position in a confidential addendum (GPR 19/1).
12 Clark, 172.
13 Ibid., 173–174. See also W.J. Brown, *The Communist Movement and Australia: An Historical Outline, 1890s to 1980s* (Haymarket: Australian Labor Movement History Publications, 1986).
14 Levy: 'You have it in your blood from your grandfather.' Letter from Levy to Pitt-Rivers, 4 May 1935 (GPR 18/2).
15 Draft letter from Pitt-Rivers to Lord Forster, 12 December 1920 (Pitt-Rivers Private Papers).
16 George Lane Fox Pitt-Rivers, 'Aua Island: Ethnographical and Sociological Features of a South Sea Pagan Society', *The Journal of the Royal*

Anthropological Institute of Great Britain and Ireland 55 (1925): 426–427;
Rainer F. Buschmann, *Anthropology's Global Histories: The Ethnographic
Frontier in German New Guinea, 1870–1935*, Perspectives on the Global Past
(Honolulu: University of Hawaii Press, 2009), 41.

17 Pitt-Rivers, 'Aua Island: Ethnographical and Sociological Features of a South
Sea Pagan Society', 427; Buschmann, 41.

18 Buschmann, 47.

19 Ibid., 125.

20 See press clipping from *Deutsche Kolonialzeitung,* 22 September 1904, relating
to 'Ermordung des händlers Reimers' (GPR 6/3).

21 GPR's account of the accident is harrowing: see GPR 6/3.

22 Buschmann, 125.

23 Pitt-Rivers, 'Aua Island: Ethnographical and Sociological Features of a South
Sea Pagan Society', 427.

24 Buschmann, 125.

25 Pitt-Rivers, 'Aua Island: Ethnographical and Sociological Features of a South
Sea Pagan Society', 426.

26 Ibid.

27 Michael W. Young, *Malinowski: Odyssey of an Anthropologist, 1884–1920*
(New Haven; London: Yale University Press, 2004), 159–160.

28 Ibid., 160.

29 Ibid., 161.

30 Letter from Lt. Governor of Papua to the Minister of Home and Territories,
Canberra, 17 December 1927 (National Archives of Australia, Department of
Home Affairs Papers, A1/ 44424).

31 Ibid.

32 See, for instance, G. Pitt-Rivers, 'The Effect on Native Races of Contact with
European Civilisation', *Man* 27 (1927): 2–10; George Henry Lane Fox Pitt-
Rivers, *The Clash of Culture and the Contact of Races: An Anthropological
and Psychological Study of the Laws of Racial Adaptability,with Special
Reference to the Depopulation of the Pacific and the Government of
Subject Races* (London: George Routledge, 1927); Pitt-Rivers, 'Aua Island:
Ethnographical and Sociological Features of a South Sea Pagan Society'.

33 Pitt-Rivers, xii.

34 See draft of 'methodology' document, GPR 27/1.

35 GPR 8: 'Outside the camphouse in Mekeo, Southern Papua' and other
photographs.

36 George Pitt-Rivers, *The Story of the Ancient Manor of Hinton St Mary*
(Farnham: The Pitt-Rivers Museum, 1947), 29.

37 Ibid.

38 Letters from Henry Balfour to Pitt-Rivers, 8 March 1923, 21 February 1924
(GPR 21/1).

39 Letter from Henry Balfour to Pitt-Rivers, 8 March 1923 (GPR 21/1).

40 Press clipping from 'Mr. W. Rodier' (undated) (GPR 37/1).

41 Press clipping 'Killing the Natives with Kindness', *Steads,* 13 May 1922
(GPR 37/1).

42 Press clipping from *The Medical Journal of Australia,* 10 May 1923 (GPR 37/1).

43 Press clipping 'An Interesting Lecture', *The Australasian,* March 1922
(GPR 37/1).

44 Press clipping 'Killing the Natives with Kindness', *Steads,* 13 May 1922 (GPR 37/1).

45 Ibid. This statement was almost certainly influenced by his reading of Malinowski's *Argonauts*, which he would have been reading at the time this lecture was given.

46 Press clippings related to Pan-Pacific Science Congress, 'Who's Who at the Congress' (GPR 37/1).

47 Young, 450. See correspondence from Masson to Pitt-Rivers, May–July 1922 (GPR 21/1).

48 Letter from Masson to Pitt-Rivers, 3 July 1922 (GPR 21/1).

49 Young, 450.

50 Ibid., 165–167.

51 Ibid., 165–173.

52 Ibid., 265–266.

53 George W. Stocking, *After Tylor: British Social Anthropology, 1888–1951* (Madison: University of Wisconsin Press, 1995), 253–255.

54 Ibid., 253–254.

55 Helena Wayne, ed. *The Story of a Marriage: The Letters of Bronislaw Malinowski and Elsie Masson*, 2 vols., vol. 2 (London: Routledge, 1995), 3.

56 Ibid., 14.

57 Bronislaw Malinowski, *Argonauts of the Western Pacific: An Account of Native Enterprise and Adventure in the Archipelagoes of Melanesian New Guinea* (London, New York: George Routledge & Sons, Ltd., 1922).

58 Stocking, 273. The fact that Malinowski's accounts of his fieldwork were often exaggerated, as shown by his diaries from the period, caused a large amount of reflection within the field in the 1960s when they appeared in print. See Young, 372; Stocking, 272.

59 Cf. Stocking, 272.

60 Ibid., 273.

61 Pitt-Rivers, xiv; C.G. Seligman, F.R. Barton and E.L. Giblin, *The Melanesians of British New Guinea* (Cambridge: The University Press, 1910).

62 Letter from Seligman to Pitt-Rivers, 2 October 1922 (GPR 19/5).

63 Letter from D.D. Braham, editor of *The Forum*, 31 August 1922 (GPR 21/1).

64 Letter from Malinowski to Pitt-Rivers, 12 December 1923 (GPR 19/3).

65 Ibid.

66 Young, 332–333.

67 Ibid., 243; Stocking, 246–247.

68 Letter from Levy to Pitt-Rivers, 25 November 1923 (GPR 19/2).

69 Stocking, 291.

70 There has been much written on functionalism: see, for instance, Ian Charles Jarvie, *Functionalism*, Basic Concepts in Anthropology (Minneapolis: Burgess Publishing Company, 1973); Andrzej K. Paluch, *Malinowski*, Wyd. 2. ed. (Warszawa: Wiedza Powszechna, 1983); George W. Stocking, *Malinowski, Rivers, Benedict and Others: Essays on Culture and Personality* (Madison: University of Wisconsin Press, 1986); Michael W. Young, 'The Careless Collector: Malinowski and the Antiquarians', in *Hunting the Gatherers: Ethnographic Collectors, Agents and Agency in Melanesia, 1870s-1930s*, ed. Michael O'Hanlon and Robert Louis Welsch (New York; Oxford: Berghahn Books, 2000); Young.

71   Cf. Young, 575.
72   Ibid., 455–456.
73   See correspondence related to appointments, April 1923 (GPR 21/1).
74   See, for instance, correspondence with Ministry of Health, 6 August 1924 (GPR 21/1).
75   Pitt-Rivers, 'Aua Island: Ethnographical and Sociological Features of a South Sea Pagan Society'.
76   Pitt-Rivers had fortuitously received a 19-volume set of Nietzsche's complete works in English translation from Levy just before he departed Australia (Letter from Pitt-Rivers to Levy, 15 May 1925 – GPR 19/2). Pitt-Rivers, *The Story of the Ancient Manor of Hinton St Mary*, 29.
77   Letters from Pitt-Rivers to RAI, May 1925 (GPR 21/2).
78   Letter from Haddon to Pitt-Rivers, 8 October 1925 (GPR 21/2). Letter from Malinowski to Pitt-Rivers, 18 May 1925 (GPR 19/3).
79   See, for instance, Pitt-Rivers, 'The Effect on Native Races of Contact with European Civilisation'; Pitt-Rivers, 'Aua Island: Ethnographical and Sociological Features of a South Sea Pagan Society'; George Pitt-Rivers, 'Sex Ratios and Cultural Contact', *Man* 27 (1927): 59–60; George Pitt-Rivers, 'Depopulation in Melanesia', *Man* 28 (1928): 213–215; George Pitt-Rivers, 'Papuan Criminals and British Justice', *Man* 29 (1929): 21–22.
80   Elazar Barkan, *The Retreat of Scientific Racism: Changing Concepts of Race in Britain and the United States between the World Wars* (Cambridge: Cambridge University Press,1992), 42. See also Jonathan Sawday, '"New Men, Strange Faces, Other Minds": Arthur Keith, Race and the Piltdown Affair (1912–53)', in *Race, Science and Medicine, 1700–1960*, ed. Waltraud Ernst and Bernard Harris (London: Routledge, 1999).
81   There is much historiography on Piltdown Man: see Ronald Millar, *The Piltdown Mystery: The Story Behind the World's Greatest Archaeological Hoax* (Seaford: S.B., 1998); Ronald William Millar, *The Piltdown Men* (London: Gollancz, 1972); Ronald William Millar, *The Piltdown Men: [a Case of Archaeological Fraud]* (St. Albans: Paladin Frogmore, 1974); Noel Roberts, *From Piltdown Man to Point Omega: The Evolutionary Theory of Teilhard De Chardin*, Studies in European Thought (New York; Oxford: P. Lang, 2000); Miles Russell, *The Piltdown Man Hoax: Case Closed* (Stroud: History, 2012); Frank Spencer and British Museum (Natural History), *The Piltdown Papers 1908–1955: The Correspondence and Other Documents Relating to the Piltdown Forgery* (London: Oxford University Press: British Museum (Natural History), 1990); Frank Spencer and Ian Langham, *Piltdown: A Scientific Forgery* (London: Natural History Museum Publications, 1990); Francis Vere, *The Piltdown Fantasy* (London: Cassell, 1955); J.S. Weiner, *The Piltdown Forgery* (London; New York: Oxford University Press, 1955).
82   Arthur Keith, *An Autobiography* (London: Watts, 1950), 326–330.
83   David Waterston, 'The Piltdown Mandible', *Nature* 92, no. 2298 (1913): 319.
84   Keith has been mentioned as a possible suspect: see Russell; Sawday; Weiner.
85   Arthur Keith, *The Place of Prejudice in Modern Civilization (Prejudice and Politics): Being the Substance of a Rectorial Address to the Students of Aberdeen University* (London: Williams & Norgate Ltd., 1931), 34.
86   Ibid., 35.

87   Barkan, 288.

88   Arthur Keith, 'Introduction', in *Weeds in the Garden of Marriage*, ed. George Henry Lane Fox Pitt-Rivers (London: N. Douglas, 1931).

89   Ibid., ix–x.

90   Unpublished autobiography of Catherine D. Taylor, 'From University to War' chapter, p. 41 (Private Collection).

91   Pitt-Rivers, ix.

92   Pitt-Rivers referred to the incident as an 'unfortunate accident' Ibid., xii.

93   Ibid.

94   Stocking, *After Tylor: British Social Anthropology, 1888–1951*, 394.

95   Pitt-Rivers, 240.

96   Ibid., 3.

97   Ibid., 13.

98   Ibid., 122–127.

99   Ibid., 131–132. Malinowski observed that most native societies he encountered denied a link between male insemination and pregnancy, which perhaps presents one origin for Pitt-Rivers' differing but similar views: see Young, 386–387.

100   Pitt-Rivers, 239.

101   Ibid., 239–240.

102   Ibid., 241.

103   Stocking, *After Tylor: British Social Anthropology, 1888–1951*, 395–396.

104   This view made Malinowski a number of powerful enemies in the political establishment: see Young, 380–381.

105   Letter from Wyndham Lewis to Pitt-Rivers, 7 May 1937 (GPR 18/2).

106   Sander L. Gilman, 'Fat as Disability: The Case of the Jews', *Literature and Medicine* 23, no. 1 (2004): 50–51.

107   Pitt-Rivers, 'The Effect on Native Races of Contact with European Civilisation', 3.

108   Ibid.

109   Ibid., 9–10.

110   Letter to George Pitt-Rivers from Tony Fowlitt, Rhodesia, 30 January 1928 (GPR 21/2).

111   John W. Cooper, 'The Preservation of Native Races', *Journal of Heredity* 19, no. 5 (1928): 234.

112   Wayne, ed., 49, 85, 93.

113   See a letters from Malinowski to his wife dated 13 and 14 October 1927 in which he recounts asking Pitt-Rivers to serve as an examiner and arguing with him over the role of Christianity in society (Bronislaw Malinowski Papers, London School of Economics and Political Science Archive, 34/26).

114   Raymond Firth, *Primitive Economics of the New Zealand Maori* (London: G. Routledge, 1929), 467n. See also Bronislaw Malinowski, Elsie Masson and Helena Wayne, *The Story of a Marriage: The Letters of Bronislaw Malinowski and Elsie Masson*, 2 vols., vol. 2 (London: Routledge, 1995), 101.

115   Julian Huxley to Pitt-Rivers, 24 June 1927 (GPR 18/4). Interestingly, Pitt-Rivers' increasingly acrimonious handwritten replies survive in the Huxley archive in Texas. See letters from 20 April and 15 August 1927 (HUX).

116   Letter from Home and Territories Department to Lieutenant Governor of Papua, 17 December 1927 (National Archives of Australia, Department of Home Affairs Papers, A1/ 44424).

117 Ibid.
118 Press clipping, *The Sunday Times (Sydney)*, 1 January 1928 (National Archives of Australia, Department of Home Affairs Papers, A1/ 44424).
119 Letter from Murray to Home and Territories Department, 9 February 1928 (National Archives of Australia, Department of Home Affairs Papers, A1/ 44424).
120 Wayne, ed., 104.
121 Ibid., 101. Also see original letter from Malinowski to his wife, 14 October 1927 (Bronislaw Malinowski Papers, London School of Economics and Political Science Archive 34/26).
122 The history of fascism has been extensively considered by historians: see, for instance, Philip Coupland, 'H.G. Wells's "Liberal Fascism"', *Journal of Contemporary History* 35, no. 4 (2000): 541–558; Stephen Dorril, *Blackshirt: Sir Oswald Mosley and British Fascism* (London; New York: Penguin Books, 2007); Martin Pugh, '*Hurrah for the Blackshirts!': Fascists and Fascism in Britain between the Wars* (London: Jonathan Cape, 2005); Dan Stone, 'The English Mistery, the BUF, and the Dilemmas of British Fascism', *The Journal of Modern History* 75, no. 2 (2003): 336–358.
123 Letter from Levy to Pitt-Rivers, 2 April 1924 (GPR 19/2).
124 Letter from Levy to Pitt-Rivers, 20 July 1924 (GPR 19/2).
125 Oscar Levy, 'Mussolini Says: "Live Dangerously": Italian Premier and Leader of Il Fascismo Acknowledges Nietzsche As Spritual Master and Avows That He Has Tried to Follow His Precept', *The New York Times*, 9 November 1924.
126 'The Social and Political Revolution in Italy' by George Pitt-Rivers, 1920s (GPR 11/1).
127 Ibid.
128 See letters between Pitt-Rivers and the Home Office, October–December 1925 (HO 382/93/1).
129 See late 1925/early 1926 letters (GPR 19/2).

Chapter Four

1 Letter from Leonard Dudley Buxton to Pitt-Rivers, 23 September 1930 (GPR 20/5): 'I find it difficult to advise you about an Oxford D.Sc., this degree is not easy to get, it has purposely been made even more difficult in recent years since the inauguration of the new doctorates and the University is jealous in conferring it. I think that a single book would not be considered sufficient by the board, usually a series have been sent in. In many cases of course no books are submitted but papers in scientific periodicals, which may often represent considerably more research than a book, and may contain greater contributions to knowledge which is the real test of the degree. In any case I should put in, if I were you, all your most important work, especially that which embodies original research.'
2 Helena Wayne, ed. *The Story of a Marriage: The Letters of Bronislaw Malinowski and Elsie Masson*, 2 vols., vol. 2 (London: Routledge, 1995), 134.
3 Letter from Malinowski to Robert H. Lowie, 22 March 1929 (Robert H. Lowie papers, Bancroft Library, University of California, Berkeley (C-B 927, Box 12)).

4   Letter from Lowie to Pitt-Rivers, 24 January 1928 (GPR 17/2): 'A trip
     East, from which I have only returned a short time ago, has delayed my
     acknowledgement of your kindness in sending me your book on "The Clash
     of Culture and the Contact of Races". Permit me to offer my belated thanks.
     It has naturally been of great interest to me, and I am particularly intrigued
     by your observations on the prevalence of polygamy. Your treatment
     certainly ought to stimulate much intensive field-work on the subject, so far
     as conditions still permit. Dealing as you do with a great variety of topics,
     you are doubtless prepared for disagreement on some of them, and I confess
     to not always finding myself in complete harmony with your views. That,
     however, is a minor matter. The book is undoubtedly stimulating, and I hope
     you will soon allow us to have a fuller account of your Oceanism researches'.
5   Letter from Pitt-Rivers to Lowie, 30 May 1929 (Lowie Papers, Box 13).
6   E.E. Evans-Pritchard, 'Foreword', in *The People of the Sierra*, ed. Julian
     Alfred Pitt-Rivers (London; Beccles: William Clowes and Sons, Ltd.,
     1954), ix.
7   Letter from Pitt-Rivers to Julian Pitt-Rivers, 12 July 1948 (Pitt-Rivers Private
     Papers).
8   Peter Rivière, *A History of Oxford Anthropology*, Methodology and History
     in Anthropology (New York: Berghahn Books, 2007), 136.
9   Peter Davis, 'How All Souls Got Its Anthropologist', in *A History of Oxford
     Anthropology*, ed. Peter Rivière (New York: Berghahn Books, 2007), 73.
10   Letter from Pitt-Rivers to Julian Pitt-Rivers, 12 July 1948 (Pitt-Rivers Private
     Papers). Pitt-Rivers was particularly insulted that Radcliffe-Brown was
     teaching his son the 'meaning' of *The Clash of Culture*.
11   Letter from Pitt-Rivers to Sir Cecil Hanbury MP, 6 June 1935 (GPR 20/1).
12   Letter from Pitt-Rivers to Malinowski, 4 September 1927 (Malinowski
     Papers, MS 19, Yale University Library Manuscripts and Archives, Series I,
     Box 6, Folder 504).
13   George Pitt-Rivers, *The Story of the Ancient Manor of Hinton St Mary*
     (Farnham: The Pitt-Rivers Museum, 1947), 50.
14   Malcolm Muggeridge, *Chronicles of Wasted Time: The Infernal Grove*, vol. 2
     (London: Fontana, 1975), 126.
15   Court minutes, Pitt-Rivers vs. Pitt-Rivers, 3 April 1929–15 July 1929
     (J 77.2644).
16   Ibid., January 1930 (J 77.2644).
17   'Mary Hinton (I) (1896–1979)', IMDB, http://www.imdb.com/name/
     nm0386003/ (accessed 8 June 2014). For an example of her theatre work,
     see 'The Theatres', *The Times*, 6 September 1928.
18   J.R. Tata, 'Rosalind Venetia Pitt-Rivers', The Royal Society, http://thyroid.
     org/wp-content/uploads/2012/04/Rosalind_Venetia_Pitt-Rivers.pdf (accessed
     13 June 2014).
19   Michael Brock, 'Stanley, (Beatrice) Venetia (1887–1948)', Oxford Dictionary
     of National Biography, http://www.oxforddnb.com/view/article/41069
     (accessed 13 June 2014). See also H.H. Asquith, M.G. Brock and Eleanor
     Brock, *Letters to Venetia Stanley* (Oxford: Oxford University Press, 1985);
     Naomi B. Levine, *Politics, Religion, and Love: The Story of H.H. Asquith,
     Venetia Stanley and Edwin Montagu* (New York; London: New York
     University Press, 1991).

20   Ibid., 328. There is extensive writing on the Mitford family: for major works, see Leslie Brody, *Irrepressible: The Life and Times of Jessica Mitford* (Washington, DC; London: Counterpoint, 2011); Jonathan Guinness and Catherine Guinness, *The House of Mitford* (London: Fontana, 1985); Selina Hastings, *Nancy Mitford: A Biography*, Vintage Lives (London: Vintage, 2002); Mary S. Lovell, *The Mitford Girls: The Biography of an Extraordinary Family* (London: Little, Brown, 2001); Jessica Mitford and Peter Y. Sussman, *Decca: The Letters of Jessica Mitford* (London: Weidenfeld & Nicolson, 2006); Nancy Mitford, Heywood Hill and John Saumarez Smith, *The Bookshop at 10 Curzon Street: Letters between Nancy Mitford and Heywood Hill 1952–73* (London: Frances Lincoln, 2004); Nancy Mitford and Charlotte Mosley, *The Letters of Nancy Mitford: Love from Nancy* (London: Hodder & Stoughton, 1993); Nancy Mitford, Evelyn Waugh and Charlotte Mosley, *The Letters of Nancy Mitford and Evelyn Waugh*, Penguin Modern Classics (London: Penguin, 2010); David Rehak, *Hitler's English Girlfriend: The Story of Unity Mitford* (Stroud: Amberley, 2011); Laura Thompson, *Life in a Cold Climate: Nancy Mitford: A Portrait of a Contradictory Woman* (London: Review, 2003).

21   Jamshed R. Tata, 'Rivers, Rosalind Venetia Lane Fox Pitt- (1907–1990)', *Oxford Dictionary of National Biography*, http://www.oxforddnb.com/view/article/57570 (accessed 13 June 2014); Tata.

22   Tata.

23   Ibid., 343–344.

24   Francis Galton, *Hereditary Genius: An Inquiry into Its Laws and Consequences* (London: Macmillan, 1869), 1.

25   Eugenics has generated a large historiography of its own: for major works, see, for instance, G.R. Searle, *Eugenics and Politics in Britain, 1900–1914* (Leyden: Noordhoff International Publishing, 1976); G.R. Searle, 'Eugenics and Class', in *Biology, Medicine and Society, 1840–1940*, ed. Charles Webster (Cambridge: Cambridge University Press, 1981); Marius Turda and Paul Weindling, 'Eugenics, Race and Nation in Central and Southeastern Europe, 1900-1940: A Historiographic Overview', in *"Blood and Homeland": Eugenics and Racial Nationalism in Central and Southeast Europe, 1900–1940*, ed. Marius Turda and Paul Weindling (Budapest: Central European University Press, 2007); Peter Weingart, 'German Eugenics between Science and Politics', *Osiris 5*, no. 1 (1989): 260; John Welshman, 'Eugenics and Public Health in Britain, 1900-40: Scenes from Provincial Life', *Urban History* 24, no. 1 (1997): 57–75; Jan A. Witkowski and John. R. Inglis, eds., *Davenport's Dream: 21st Century Reflections on Heredity and Eugenics* (Cold Spring Harbor, NY: Cold Spring Harbor Laboratory Press, 2008); Jeremy Noakes, 'Nazism and Eugenics: The Background to the Nazi Sterilization Law of 14 July 1933', in *Ideas into Politics: Aspects of European History 1880–1950*, ed. Roger Bullen, H.Pogge von Strandmann and Antony Polonsky (London: Croom Helm, 1984); Diane Paul, 'Eugenics and the Left', *Journal of the History of Ideas* 45, no. 4 (1984): 567–590; Garland E. Allen, 'The Eugenics Record Office at Cold Spring Harbor, 1910-1940: An Essay in Institutional History', *Osiris* 2, (1986): 225–264; John Macnicol, 'Eugenics and the Campaign for Voluntary Sterilization in Britain between the Wars', *Social History of Medicine* 2, no. 2 (1989): 147–169; Stefan Kühl, *The Nazi Connection: Eugenics, American*

Racism, and German National Socialism (New York; Oxford: Oxford University Press, 1994); Daniel J. Kevles, *In the Name of Eugenics: Genetics and the Uses of Human Heredity* (Cambridge, MA: Harvard University Press, 1995); Edwin Black, *War against the Weak: Eugenics and America's Campaign to Create a Master Race* (New York; London: Four Walls Eight Windows, 2003); Lucy Bland, 'British Eugenics and "Race Crossing": A Study of an Interwar Investigation', *New Formations* 60, (2006): 66–78; Nathaniel C. Comfort, 'Jan A. Witkowski; John R. Inglis (Editors). Davenport's Dream: Twenty-First Century Reflections on Heredity and Eugenics', *Isis* 100, no. 1 (2009): 191–192; Alison Bashford and Philippa Levine, *The Oxford Handbook of the History of Eugenics* (New York; Oxford: Oxford University Press, 2010); Marius Turda, *Modernism and Eugenics* (New York: Palgrave Macmillan, 2010); Stefan Kühl, *For the Betterment of the Race: The Rise and Fall of the International Movement for Eugenics and Racial Hygiene* (New York, NY: Palgrave Macmillan, 2013); Richard A. Soloway, *Demography and Degeneration: Eugenics and the Declining Birthrate in Twentieth-Century Britain* (Chapel Hill: University of North Carolina Press, 1990); Nathaniel C. Comfort, *The Science of Human Perfection: How Genes Became the Heart of American Medicine* (New Haven: Yale University Press, 2012).

26 Soloway, 38–44.

27 Ibid., 60.

28 Ibid., 171; Ernest James Lidbetter, *Heredity and the Social Problem Group* (London: E. Arnold, 1933).

29 California's eugenicists were proud of their achievement in this area: see Ezra Seymour Gosney and Paul Bowman Popenoe, *Sterilization for Human Betterment; a Summary of Results of 6,000 Operations in California, 1909–1929* (New York: The Macmillan Company, 1931); Paul Popenoe, 'Public Opinion on Sterilization in California', *Eugenical News* XX, no. 5 (1935): 73–75. See also Cora B.S. Hodson, *Human Sterilization to-Day: A Survey of the Present Position* (London: Watts & Co., 1934). California's eugenic sterilization movement has rightly received extensive historiographical attention: see Kevles, 114–115; Alexandra Stern, *Eugenic Nation: Faults and Frontiers of Better Breeding in Modern America* (Berkeley: University of California Press, 2005); Black.

30 Kevles, 111; Paul A. Lombardo, *Three Generations, No Imbeciles: Eugenics, the Supreme Court, and Buck V. Bell* (Baltimore; London: Johns Hopkins University Press, 2010).

31 Anonymous, 'Sterilization Laws: More Than One-Half of the States Have Eugenical Sterilization Laws', *Eugenical News* XVI, no. 8 (1931): 129.

32 Kevles, 59–60.

33 Soloway, 33.

34 Letter from Pitt-Rivers to Ferdinand C.S. Schiller, 14 April 1926 (F.C.S. Schiller Papers, University of California, Los Angeles Library Special Collections, Box 2).

35 T.R. Malthus, *An Essay on the Principle of Population, as It Affects the Future Improvement of Society. With Remarks on the Speculations of Mr. Godwin, M. Condorcet, and Other Writers* (London: Printed for J. Johnson, 1798).

36 Soloway, 4–5.

37 Ibid., 18.

38 George Henry Lane Fox Pitt-Rivers, *The Clash of Culture and the Contact of Races: An Anthropological and Psychological Study of the Laws of Racial Adaptability, with Special Reference to the Depopulation of the Pacific and the Government of Subject Races* (London: George Routledge, 1927), 30–35.

39 See Francis Galton, *Essays in Eugenics* (London: Eugenics Education Society, 1909).

40 Kevles, 21.

41 Ellis' groundbreaking work, and its influence on psychoanalysis, have been widely discussed: see Arthur Calder-Marshall, *Havelock Ellis: A Biography* (London: R. Hart-Davis, 1959); John Stewart Collis, *An Artist of Life: A Study of the Life and Work of Havelock Ellis* (London: Cassell, 1959); Ivan Crozier, 'Havelock Ellis, Eugenicist', *Studies in History and Philosophy of Science Part C: Studies in History and Philosophy of Biological and Biomedical Sciences* 39, no. 2 (2008): 187–194; Rose Freeman-Ishill, *Havelock Ellis* (Berkeley Heights: Oriole Press, 1942); Chris Nottingham, *The Pursuit of Serenity: Havelock Ellis and the New Politics* (Amsterdam: Amsterdam University Press, 1999); Houston Peterson, *Havelock Ellis, Philosopher of Love* (Boston; New York: Houghton Mifflin Company, 1928); Paul A. Robinson, *The Modernization of Sex: Havelock Ellis, Alfred Kinsey, William Masters, and Virginia Johnson* (Ithaca: Cornell University Press, 1989); Sheila Rowbotham and Jeffrey Weeks, *Socialism and the New Life: The Personal and Sexual Politics of Edward Carpenter and Havelock Ellis* (London: Pluto Press, 1977). Ellis' views on eugenics and homosexuality presented in Crozier, 'Havelock Ellis', 173.

42 Letter from Havelock Ellis to George Pitt-Rivers, 11 December 1919 (GPR 18/3).

43 Pitt-Rivers, 31.

44 Letter to the Eugenics Education Society from GPR, 29 April 1920 (GPR 21/2).

45 'Reply to the Council of E.E.S', *The Eugenics Review* XII, no. 1 (1920–1921): 73.

46 Soloway, 87–88.

47 Ibid., 88–89.

48 Ibid., 88.

49 Draft article: 'Birth Control' by L. Darwin (C.P. Blacker Papers, CPB/B.1/5).

50 See Gavin Schaffer, '"Scientific' Racism Again?": Reginald Gates, the Mankind Quarterly and the Question of "Race" in Science after the Second World War', *Journal of American Studies* 41, no. 2 (2007): 253–278.

51 Soloway, 178.

52 This strange affair was well examined by June Rose, *Marie Stopes and the Sexual Revolution* (Stroud: Tempus, 2007), 76–78. On Stopes, see also Keith Briant, *Marie Stopes, a Biography* (London: Hogarth Press, 1962); Ruth E. Hall, *Marie Stopes: A Biography* (London: Deutsch, 1977); Ruth E. Hall, *Dear Dr Stopes: Sex in the 1920s* (Harmondsworth: Penguin, 1981); Aylmer Maude, *The Authorized Life of Marie C. Stopes* (London: Williams & Norgate Ltd., 1924); Robert A. Peel, *Marie Stopes, Eugenics and the English Birth Control Movement: Proceedings of a Conference Organised by the Galton Institute, London, 1996* (London: Galton Institute, 1997).

53 Marie Carmichael Stopes, *Married Love: A New Contribution to the Solution of Sex Difficulties*, Seventh ed. (London: G. P. Putnam's Sons, 1919).

54 Clementine Churchill to Winston S. Churchill, 11 August 1918 (Chartwell Papers 1/125/9-11).

55 Soloway, 178–179.

56 Richard Overy, *The Morbid Age: Britain between the Wars* (London: Allen Lane, 2009), 99.

57 See, for instance, Anonymous, 'Mental Deficients', *The Birth Control News* VIII, no. 1 (1929): 2.

58 Soloway, 179–180.

59 'Eugenics – Or a Return to Savagery?' by George Pitt-Rivers, *The Herald* (Australia?), Thursday evening, 22 November 1923 (GPR 6/1).

60 Society for Constructive Birth Control and Racial Progress address, Essex Hall, 25 October 1932 (GPR 6/2).

61 See correspondence between Stopes and Pitt-Rivers, early 1930s (GPR 27/3).

62 Letter from Pitt-Rivers to Stopes, 9 March 1934 (GPR 20/1).

63 George Pitt-Rivers, 'Correspondence', *The Eugenics Review* XII, (1920–1921): 71–72.

64 Eugenics Society Council Minutes, December 1926–January 1930 (Eugenics Society Papers, Wellcome Library (hereafter EUG) L.8).

65 For more on the IFEO see, Stefan Kühl, *Die Internationale Der Rassisten: Aufstieg Und Niedergang Der Internationalen Bewegung Für Eugenik Und Rassenhygiene Im 20. Jahrhundert* (Frankfurt: Campus Verlag, 1997); Stefan Kühl, 'The Relationship between Eugenics and the So-Called "Euthanasia Action" in Nazi Germany: A Eugenically Motivated Peace Policy and the Killing of the Mentally Handicapped During the Second World War', in *Science in the Third Reich*, ed. Margit Szöllösi-Janze (Oxford: Berg, 2001); Kühl, *For the Betterment of the Race: The Rise and Fall of the International Movement for Eugenics and Racial Hygiene*.

66 24 July 1929 Eugenics Society Council Minutes (EUG/L.8).

67 International Federation of Eugenic Organisations Minutes of Business Meeting, 27 September 1929, Rome (Harry H. Laughlin Papers, Truman State University C2-4-1:6).

68 Letter from Fischer to Davenport, 19 July 1929 (Max Planck Gesellschaft archive (MPG-Archiv), I Ab., Rep. 3, Num. 23).

69 Letter from Fischer to Hodson, 27 August 1930 (MPG-Archiv, I Ab., Rep. 3, Num. 23).

70 Cora B.S. Hodson, ed., *Report of the Ninth Conference of the International Federation of Eugenic Organizations* (Farnham: I.F.E.O., 1930).

71 Press clipping, 'The Future of Mankind: Sir Arthur Keith and Eugenic Control' from the Liverpool Post, 13 September 1930, From A Special Correspondent (Reginald Ruggles Gates Papers, Kings College London Archive 9/1).

72 Ibid.

73 Hodson, *Report of the Ninth Conference of the International Federation of Eugenic Organizations*, 21–26.

74 'The Urgency of Eugenic Reform: Address to the International Federation of Eugenic Organizations, 1930, Hinton St Mary, Dorset', 6–7 (Arthur Keith Papers, Royal College of Surgeons Library MS0018/2/10/13).

75 Richard A. Soloway, 'Blacker, Carlos Paton (1895–1975)', *Oxford Dictionary of National Biography* (2004). http://www.oxforddnb.com/view/article/47726 (accessed 26 July 2010). See Blacker's war

memoirs: C.P. Blacker, *Have You Forgotten Yet? The First World War Memoirs of C.P. Blacker*, ed. J.G.C. Blacker (London: Leo Cooper, 2000).

76   Soloway.

77   Robert Olby, 'Huxley, Sir Julian Sorell (1887–1975)', *Oxford Dictionary of National Biography* (2004). http://www.oxforddnb.com/view/article/31271 (accessed 20 April 2010).

78   Blacker later stated that it was left-wing zoologist Lancelot Hogben who had changed Huxley's views, of which Hogben was proud: see Paul Gary Werskey, 'British Scientists and "Outsider" Politics, 1931–1945', *Science Studies* 1, no. 1 (1971): 73.

79   Anonymous, 'Obituary: Carlos Paton Blacker', *The Lancet* 305, no. 7915 (1975): 1096.

80   C.P. Blacker, *Birth Control and the State; a Plea and a Forecast*, To-Day and to-Morrow (London: K. Paul, Trench, 1926); Kevles, 349.

81   Soloway, 186.

82   Letter from Blacker to Darwin (unsent), 30 July 1936 (CPB/B.1/5).

83   Ibid.

84   Kevles, 172.

85   Letter from Darwin to Blacker, 17 October 1930 (Darwin Family Papers, University Library, Cambridge 9368.1: 16774; also CPB/B.3).

86   Letter from Pitt-Rivers to RRG, 6 January 1933, Reginald Ruggles Gates Papers, Kings College London Archive, Correspondence 1933–1935 (7/8).

87   George Henry Lane Fox Pitt-Rivers, *Weeds in the Garden of Marriage* (London: N. Douglas, 1931).

88   Letter from T.S. Eliot, Faber & Faber, to Pitt-Rivers, 8 April 1931 (GPR 6/1).

89   Pitt-Rivers, *Weeds in the Garden of Marriage*, 4–5.

90   Ibid., 5.

91   Ibid., 14, 16–17. In 1922, Chesterton had written a polemic of his own attacking eugenics: G.K. Chesterton, *Eugenics and Other Evils* (London; New York: Cassell, 1922).

92   Pitt-Rivers, *Weeds in the Garden of Marriage*, 23.

93   Ibid., 33–34.

94   Robert Harry Lowie, *Are We Civilized?: Human Culture in Perspective* (New York: Harcourt, Brace and Company, 1929), 46; Pitt-Rivers, *Weeds in the Garden of Marriage*.

95   Lowie, 286–287.

96   Pitt-Rivers, *Weeds in the Garden of Marriage*, 47; Lowie, 281.

97   Pitt-Rivers, *Weeds in the Garden of Marriage*, 49–53.

98   Ibid., 49.

99   Ibid., 52–53.

100   Ibid., 64.

101   Ibid., 78–83.

102   Letters from Pitt-Rivers to Blacker, 8 December 1931, 4 April 1932 (EUG/C.273).

103   Letter from R.A. Fischer to Pitt-Rivers, 19 March 1932 (GPR 20/5).

104   Letter from Pitt-Rivers to Lowie, 31 December 1931 (Lowie Papers Box 13).

105   S.J. Holmes, 'In Defense of Eugenics', *Journal of Heredity* 23, no. 6 (1932): 243–244.

106   'Weeds in the Garden of Marriage', *Nature* 129, no. 3257 (1932): 493.

107 Letter dated 6 November 1931 from C.P. Blacker to Julian Huxley (Julian Huxley Papers, Rice University Library, C.P. Blacker correspondence file).

108 Letter dated 3 February 1932 from C.P. Blacker to Julian Huxley (Julian Huxley Papers, Rice University Library, C.P. Blacker correspondence file).

109 Anonymous, 'Eugenics and Religion: Summary of a Debate Arranged by the Eugenics Society (25 April 1933)', *The Eugenics Review* XXV (1933–1934): 101.

110 19 March 1932 memorandum from G. Pitt-Rivers to the Council (EUG/L.9).

111 Letter from Ellis to Pitt-Rivers, 29 March 1932, with accompanying memorandum (GPR 18/3). Paragraphing dissolved.

112 Letter from Inge to George Pitt-Rivers, 22 March 1932 (GPR 18/4).

113 Letter from Blacker to Pitt-Rivers, 21 March 1932 (EUG/C.273). Letter from Bernard Mallet to Pitt-Rivers, 22 March 1932 (GPR 20/5).

114 Letter from Bernard Mallet to Pitt-Rivers, 22 March 1932 (GPR 20/5).

115 22 April 1932 memorandum from Pitt-Rivers to the Council (EUG/C.273).

116 George Pitt-Rivers, 'Anthropological Approach to Ethnogenics: A New Perspective', *Human Biology* 4, no. 2 (1932): 251. See also George Pitt-Rivers, 'Anthropological Approaches to Ethnogenics', in *Essays Presented to C. G. Seligman*, ed. E.E. Evans-Pritchard et al. (London: K. Paul Trench Trubner & Co. Ltd., 1934).

117 Pitt-Rivers, 'Anthropological Approaches to Ethnogenics', 251. The previous version of the paragraph was the same through the phrase 'collapse of civilizations and cultures' and then continued: 'I have described Science as the conscious and systematic attempt of man to control his environment, the conscious part of adaptation in his fight for survival. But by far the greatest and most imposing advances in science have been made in man's control over the physical forces of his environment; on the organic and the cultural sides he has lagged most sadly behind. Just where there is the greatest need for a science of the control of society, in its organization and composition – quantitative and qualitative – it has not been allowed or has not been able to help.' (Pitt-Rivers, 'Anthropological Approach to Ethnogenics: A New Perspective', 249.) This generic endorsement of science was replaced with his specific endorsement of fascism.

118 George Henry Lane Fox Pitt-Rivers, 'Editor's Preface', in *Problems of Population*, ed. George Henry Lane Fox Pitt-Rivers (Port Washington, NY: Kennikat Press, 1971), 31. The most extensive analysis of the organization to date is Kühl, *Die Internationale Der Rassisten: Aufstieg und Niedergang der Internationalen Bewegung für Eugenik Und Rassenhygiene im 20. Jahrhundert*.

119 'Interim Report of the Proceedings of the First General Assembly of the International Union for the Scientific Investigation of Population Problems', *Journal of the American Statistical Association* 23, no. 163 (1928): 306–317. See also F. Landis MacKellar and Bradley W. Hart, 'Captain George Henry Lane-Fox Pitt-Rivers and the Prehistory of the IUSSP', *Population and Development Review* 40, no. 4 (2014): 653–675.

120 George Henry Lane Fox Pitt-Rivers, *Problems of Population* (London: George Allen & Unwin Ltd., 1932).

121 Letter from Sir Charles Close to William Beveridge, 20 November 1932 (Wellcome Library, Population Investigation Committee Papers, C/4).

Chapter Five

1 For a more complete account, see, among others, Richard J. Evans, *The Coming of the Third Reich* (London: Penguin Books, 2003).
2 'The Fascist Movement in England' speech by Pitt-Rivers, typed MSS dated October 1934, p. 14 (GPR 11/1).
3 Letter from George Pitt-Rivers to Michael Pitt-Rivers, 19 August 1936 (Pitt-Rivers Private Papers).
4 Letter from Malinowski to Pitt-Rivers, 22 February 1935 (GPR 18/2). See also correspondence related to Hinton St Mary visit found in Helena Wayne, ed. *The Story of a Marriage: The Letters of Bronislaw Malinowski and Elsie Masson*, 2 vols., vol. 2 (London: Routledge, 1995), 172, 174.
5 See correspondence between Pitt-Rivers and Malinowski, 1935 (GPR 18/2).
6 The last correspondence between the two appear to have been written in 1935 (GPR 18/2).
7 Dan Stone, 'Nazism as Modern Magic: Bronislaw Malinowski's Political Anthropology', *History and Anthropology* 14 (2003): 203–218.
8 A.H. Cosway, *Guide to the Tithe Act, 1936* (London: Sir I. Pitman & sons, Ltd., 1937), 1.
9 Ibid., 4–5.
10 Ibid., 6.
11 Ibid., 7–8.
12 Ibid., 8.
13 Stuart Murray, *Beyond Tithing* (Carlisle: Paternoster, 2000), 182–183.
14 Juliet Gardiner, *The Thirties: An Intimate History of Britain* (London: HarperPress, 2011), 241–242.
15 George Pitt-Rivers, 'The Tithe Proposals', *The Times*, 13 March 1936, 10.
16 'Attacks on Tithe System', *The Manchester Guardian*, 11 January 1935, 7.
17 Clippings from *Dorset Daily Echo* and *Bournemouth Daily Echo*, 10–11 January 1935 (GPR 3).
18 Clipping from *Dorset County Chronicle*, 25 July 1935 (GPR 18/2).
19 Patrick Wright, *The Village That Died for England: The Strange Story of Tyneham* (London: Vintage, 1996), 164.
20 Ibid., 163–164.
21 Catherine Taylor, *If Courage Goes: My Twenty Years in South African Politics* (Johannesburg: Macmillan, 1976), 17–19.
22 Ibid., 21–22.
23 Ibid., 26.
24 Ibid., 30.
25 Ibid., 29–30.
26 27 October 1937 note in Pitt-Rivers' security service file, KV 2/831.
27 William Makepeace Thackeray, *Vanity Fair* (London: 1848). Sharpe's response to the nickname found in unpublished manuscript chapter 'From University to War' chapter, p. 43 (Private Collection).
28 Unpublished autobiography of Catherine D. Taylor, 'From University to War' chapter, p. 44 (Private Collection).
29 Ibid., p. 40 (Private Collection).
30 Ibid., p. 45 (Private Collection).

31 Case of Pitt-Rivers vs. Pitt-Rivers, 24 July 1937 filing by the respondent (J 77/3773).
32 Ibid., paragraph 2(b) (J 77/3773).
33 Ibid. (J 77/3773).
34 A.J.P. Taylor, *Beaverbrook* (London: Hamilton, 1972), 300–303.
35 Letters between Beaverbrook and Pitt-Rivers, 24 June 1930, 10 March 1933 (GPR 18/1).
36 'Programme of events' Tithe Barn, Hinton St Mary, 7 July 1934 (GPR 18/4).
37 Transcript of 8 November 1940 hearing, p. 21 (HO 45/25745).
38 Letter from Pitt-Rivers to Lord Tavistock, 5 April 1936 (GPR 25/1).
39 Wright, 164.
40 Mosley has inspired a large historiography of his own: see, for instance, Stephen Dorril, *Blackshirt: Sir Oswald Mosley and British Fascism* (London; New York: Penguin Books, 2007); Gary Love, '"What's the Big Idea?": Oswald Mosley, the British Union of Fascists and Generic Fascism', *Journal of Contemporary History* 42, no. 3 (2007): 447–468; Robert Jacob Alexander Skidelsky, *Oswald Mosley* (London; New York: Macmillan, 1975); Leonard Wise, *Mosley's Blackshirts: The Inside Story of the British Union of Fascists, 1932–1940* (London: Sanctuary, 1986). Quote from A.K. Chesterton, *Oswald Mosley, Portrait of a Leader* (London: Action Press, Ltd., 1937), 111.
41 Richard Carr has commented on the appeal of the BUF in this period: Richard Carr, *Veteran MPs and Conservative Politics in the Aftermath of the Great War: The Memory of All That* (Farnham; Burlington, VT: Ashgate Pub., 2013).
42 Chesterton, 119.
43 Dorril, 289.
44 Cf. ibid., 250.
45 North Dorset Agricultural Defence League Constitution (GPR 13/1).
46 North Dorset Agricultural Defence League: Our Five Points (GPR 13/1).
47 Letter from T.A. Butler, BUF, February 1935 (GPR 13/1). The proposed candidate was Ronald Farquharson, one of Pitt-Rivers' associates in the WADA. He wisely decided to sit the election out.
48 Letters and documents concerning BUF political organization, February 1935 (GPR 13/1).
49 Letter from Pitt-Rivers to Mosley, 8 April 1935 (GPR 13/1).
50 Letter from F.M. Box to Henry Walton, 1 May 1935; statement of understanding between Pitt-Rivers and BUF, 1935 (GPR 13/1).
51 1 May 1935 letter from Box to Pitt-Rivers with accompanying points (GPR 13/1).
52 See letters between Pitt-Rivers and Lord Beaverbrook concerning the anti-tithe movement and the BUF, 1935 (GPR 16/6).
53 Letters between Lord Beaverbrook and Pitt-Rivers, 27 April 1935, 11 May 1935 (GPR 16/6).
54 Dorril, 364.
55 Wright, 166.
56 Anonymous, 'A "Fourth Party" in Dorset: Sir C. Hanbury's Varied Opponents', *The Times*, 31 October 1935; Wright, 167.
57 Wright, 167.
58 Anonymous, 'North Dorset by-Election: The Prime Minister's Message', *The Times*, 9 July 1937, 9, Col. D.

59   Notes from interview with G. Anthony Pitt-Rivers, 31 June 2014.

60   Dorril, 364.

61   Pitt-Rivers, 10. Clippings from *Dorset County Chronicle*, 2 July 1936 (GPR 18/2), various other newspapers (GPR 3).

62   Laura Brace, *The Idea of Property in Seventeenth-Century England: Tithes and the Individual*, Politics, Culture and Society in Early Modern Britain (Manchester: Manchester University Press, 1998); Cosway; Pitt-Rivers, 10.

63   Letter from Pitt-Rivers to RRG, 30 November 1935 (Gates Papers 7/8).

64   *Race and Culture* (London: Royal Anthropological Institute of Great Britain and Ireland, Institute of Sociology, 1936), 2.

65   Bradley W. Hart, 'Science, Politics, and Prejudice: The Dynamics and Significance of British Anthropology's Failure to Confront Nazi Racial Ideology', *European History Quarterly* 43, no. 2 (2013): 306; Elazar Barkan, *The Retreat of Scientific Racism: Changing Concepts of Race in Britain and the United States between the World Wars* (Cambridge: Cambridge University Press, 1992), 289.

66   *Race and Culture*, 2.

67   Letter from Seligman to Elliot Smith, 24 July 1934 (Raymond Firth Papers, London School of Economics Library Archive, 8/2/3).

68   Letter from Seligman to Firth, 22 April 1934 (Raymond Firth Papers, London School of Economics Library Archive, 8/2/3).

69   Ibid., 25 May 1934 (Raymond Firth Papers, London School of Economics Library Archive, 8/2/3).

70   Letter from Fleure to Firth, circulated to all Race and Culture Committee members, 20 February 1935 (GPR 11/4).

71   Letter from Pitt-Rivers to A.G. Collingridge, 19 November 1934 (Pitt-Rivers Papers 23/1).

72   Benno Müller-Hill, *Murderous Science: Elimination by Scientific Selection of Jews, Gypsies, and Others, Germany, 1933–1945* (Oxford: Oxford University Press, 1988), 81.

73   Letter from Pitt-Rivers to Loeffler, 2 April 1935 (GPR 11/3).

74   'Scientific As Against Political Implications of the Aryan Question' Memorandum by Pitt-Rivers, 1935 (GPR 11/2). Italics and emphases original.

75   *Race and Culture*, 3.

76   Ibid.

77   Ibid., 17.

78   Correspondence between Pitt-Rivers and Firth, 14–15 November 1935, letter from Firth to Pitt-Rivers, 22 November 1935 (Firth Papers 8/2/3).

79   *Race and Culture*, 4.

80   Letter from Pitt-Rivers to Firth, 15 December 1935 (Firth Papers 8/2/3).

81   For further discussion of the committee's wider implications, see Hart, 'Science, Politics, and Prejudice: The Dynamics and Significance of British Anthropology's Failure to Confront Nazi Racial Ideology'; Barkan.

82   Letter from Pitt-Rivers to Karl Astel, 7 March 1936 (GPR 17/3).

83   Pitt-Rivers' papers reveal no correspondence with either man after 1936.

84   Letter from Raymond Firth to Pitt-Rivers, 29 February 1936 (GPR 11/4).

85   Pitt-Rivers to Gates, 15 December 1933 (Reginald Ruggles Gates Papers, Kings College Library 7/8): 'The definition which I used in my own University

lectures in Germany and which appeared to get general endorsement was: "*Wir verstehen unter Rasse eine biologische Gruppe, characktiesiert durch den gemeinsamen Besitz einer nicht festbestimmened Anzahl zu einem Ganzen verbundener Züge, die sie von anderen Gruppen unterschieden.*"'

86   Letter from Pitt-Rivers to Prince Otto von Bismarck, 8 January 1936 (GPR 20/6).
87   Letter from Oscar Levy to George Pitt-Rivers, 4 May 1935 (GPR 19/2).
88   See letter from Gates to Pitt-Rivers on Athenaeum Club membership, 19 October 1935 (GPR 20/2).
89   Fragmentary Handwritten MSS, undated [1930s] (GPR 11/1).
90   For an analysis of the event's importance, see F. Landis MacKellar and Bradley W. Hart, 'Captain George Henry Lane-Fox Pitt-Rivers, Nazi Population Science, and the Prehistory of the IUSSP', *Population and Development Review* 40, no. 4 (2014): 653–675.
91   Hans Harmsen and Franz Lohse, *Bevölkerungsfragen: Bericht Des Internationalen Kongresses Für Bevölkerungswissenschaft, Berlin, 26. August-1, September 1935* (Munich: J. F. Lehmann, 1936).
92   Proceedings of divorce suit, Pitt-Rivers v. Pitt-Rivers, Sharpe intervening, 1937 (J 77/3773).
93   Ibid.
94   Letter from Pitt-Rivers to Rolf Gardiner, 19 August 1936 (GPR 22/3).
95   Ibid.
96   Undated letter from Michael Pitt-Rivers to George Pitt-Rivers, August 1936 (Pitt-Rivers Private Papers).
97   Ibid., 26 August 1936 (Pitt-Rivers Private Papers). Emphasis original.
98   Letter from Pitt-Rivers to Rolf Gardiner, 19 August 1936 (GPR 22/3).
99   For more on Gardiner, see Matthew Jefferies and Mike Tyldesley, *Rolf Gardiner: Folk, Nature and Culture in Interwar Britain* (Farnham: Ashgate, 2011).
100 Letters between Gardiner and Pitt-Rivers, 1936 (GPR 14/3).
101 See Ibid., GPR 22/3.
102 'Our Own Correspondent', 'Germans and Czechs', *The Times*, 29 August 1936, 298–300; Richard Griffiths, *Fellow Travellers of the Right: British Enthusiasts for Nazi Germany: 1933–9* (London: Constable, 1980).
103 George Henry Lane Fox Pitt-Rivers, *The Czech Conspiracy: A Phase in the World-War Plot* (London: Boswell Pub. Co., 1938), 62.
104 Ibid., 42.
105 Unpublished autobiography of Catherine D. Taylor, 'From University to War' chapter, p. 49 (Private Collection).
106 Ibid.
107 Pitt-Rivers, 62.
108 Unpublished autobiography of Catherine D. Taylor, 'From University to War' chapter, p. 50 (Private Collection).
109 See Pitt-Rivers, 60–63.
110 Unpublished autobiography of Catherine D. Taylor, 'From University to War' chapter, p. 51 (Private Collection).
111 Václav Prucha, ''The Economy and the Rise and Fall of a Small Multinational State: Czechoslovakia, 1918–1992', in *Nation, State and the Economy in History*, ed. Alice Teichova and Herbert Matis (Cambridge: Cambridge University Press, 2003), 185.

112 Ibid., 185–186; Ladislav Cabada, "From Munich to the Renewal of Czechoslovakia', in *Czechoslovakia and the Czech Republic in World Politics*, ed. Ladislav Cabada and Sárka Waisová (Lanham: Lexington Books, 2011), 34.

113 Unpublished autobiography of Catherine D. Taylor, 'From University to War' chapter, p. 49 (Private Collection). Clipping from *The Daily Express*, 3 October 1936 (GPR 3).

114 Letter from Pitt-Rivers to Stephen King-Hall, 9 April 1938 (GPR 25/3).

115 Unpublished autobiography of Catherine D. Taylor, 'From University to War' chapter, p. 51 (Private Collection).

116 Pitt-Rivers, 66. Clipping from *The Daily Express*, 3 October 1936 (GPR 3).

117 Unpublished autobiography of Catherine D. Taylor, 'From University to War' chapter, p. 51 (Private Collection).

118 Robert Blake, "Churchill, Randolph Frederick Edward Spencer (1911–1968)', *Oxford Dictionary of National Biography*, http://www.oxforddnb.com/view/article/37283 (accessed 12 July 2014). See also Winston S. Churchill, *His Father's Son: The Life of Randolph Churchill* (London: Weidenfeld & Nicolson, 1996), 116–117.

119 Letter from Richard Findlay to Oswald Mosley, 22 February 1935 (Randolph Churchill papers, Churchill Archives Centre, 2/4).

120 Statement to the press by Richard Findlay, 22 February 1935 (RDCH 2/4).

121 Letter from Findlay to Warner Allen, *The Saturday Review*, 3 November 1936 (GPR 17/3).

122 Unpublished autobiography of Catherine D. Taylor, 'From University to War' chapter, p. 37 (Private Collection).

123 Ibid., p. 38 (Private Collection).

124 George Pitt-Rivers, "The Truth About Spain: A Remarkable Report', *The Anglo-German Review* 1, no. 4 (1937): 158.

125 Ibid., 157–159.

126 Letter from Pitt-Rivers to the Earl of Plymouth, 28 December 1936 (GPR 25/1).

127 Letter from Private Secretary, the Earl of Plymouth to Pitt-Rivers, 20 January 1937 (GPR 25/1).

128 Letter from Pitt-Rivers to Sir Eric Phipps, 4 December 1936 (GPR 17/3).

129 Letter from George Pitt-Rivers to Sir Arnold Wilson, 18 December 1936 (GPR 25/1).

130 The AGF has been the subject of past analysis: see, for instance, G.T. Waddington, "'An Idyllic and Unruffled Atmosphere of Complete Anglo-German Misunderstanding": Aspects of the Operations of the Dienststelle Ribbentrop in Great Britain, 1934–1938', *History* 82, no. 265 (1997): 44–72; Griffiths.

131 Waddington, 45–60.

132 Letter from Pitt-Rivers to W.[*sic*].P. Conwell-Evans, 1 February 1937 (GPR 25/2).

133 Bertrand de Jouvenal, 'Heart to Heart Talk with Hitler', *The Anglo-German Review* 1, no. 4 (1937): 1.

134 'Military Tailor', 'Letters to the Editor', *The Anglo-German Review* 1, no. 2 (1936): 96.

135 Griffiths, 184; Waddington, 62.

136 Griffiths, 146, 222. Quote from Sir Robert Vansittart Papers, Churchill Archives Centre, Cambridge, 2/42. Emphasis original.

137   Among others, Conwell-Evans passed information to Vansittart: see
      Vansittart Papers, Churchill Archives Centre, Cambridge, 2/42.
138   Richard C. Thurlow, *Fascism in Britain: A History, 1918–1985* (Oxford, UK;
      New York, NY: Blackwell, 1987), 100–101; Griffiths, 50–51. Letter from
      Pitt-Rivers to Conwell-Evans, 1 February 1937 (GPR 25/2).
139   'The Anglo-German Fellowship: List of Donations', 1935–36 Annual Report
      (GPR 25/1).
140   Cf. Thurlow, 82.
141   Griffiths, 179–182; Thurlow, 164. For more on Domvile's views, see Barry
      Domvile, *By and Large* (London: Hutchinson, 1936); Barry Domvile,
      From Admiral to Cabin Boy (London: Boswell Pub. Co., 1947). Domvile's
      detention file and accounts of his views can be found in HO 283/31.
142   Letter from Pitt-Rivers to Lord Mount Temple, 18 May 1937 (GPR 25/2).
      Pitt-Rivers alternatively suggested that the remark had been made to him
      directly at the dinner – see letter from Pitt-Rivers to Conwell-Evans, 1
      February 1937 (GPR 25/2).
143   Letter from Pitt-Rivers to Lord Mount Temple, 1 June 1937 (GPR 25/2).
144   Letter from Pitt-Rivers to Conwell Evans, 1 February 1937; Receipt for
      payment of dues to AGF by Sharpe, 15/2/37 (GPR 25/2).
145   Letter from Pitt-Rivers to Conwell-Evans, 1 February 1937 (GPR 25/2).
146   Letter from Sharpe, et al. to AGF Secretary, 21 February 1937 (GPR 25/2).
147   Letter from Wright to Sharpe, 16 February 1937 (GPR 25/2). Copies of
      Wright's official correspondence within the AGF found in the GPR 25/1-3.
      See letter terminating Wright's employment, 18 May 1937 and letter from
      Wright to Karl Budding, 25 May 1937(GPR 25/2).
148   Letter from Wright to Budding, 25 May 1937, pp. 9–13 (GPR 25/2).
149   Ibid., pp. 9–13 (GPR 25/2).
150   Ibid., p. 9 (GPR 25/2).
151   Signed note attesting to proceedings of meeting, signed by Elwin Wright,
      11 May 1938 (GPR 25/3).
152   See letter from Wright to Sir Raymond Beazley, 3 December 1937 (GPR 25/2).
153   Ibid. (GPR 25/2).
154   Frances Lonsdale Donaldson, *Edward VIII* (London: Weidenfeld &
      Nicolson, 1974), 276–278.
155   Ibid., 277.
156   Ibid., 282. See also Philip Ziegler, *King Edward VIII: The Official Biography*
      (London: Collins, 1990), 277–279.
157   Quoted in Pitt-Rivers' case summary document, 1942, KV 2/831 and also
      in 25 December 1936 letter from Pitt-Rivers to the Under-Secretary of State
      (Pitt-Rivers Private Papers).
158   Letter from Pitt-Rivers to Under Secretary of State, 20 February 1937
      (GPR 25/2). Baldwin had remarked in parliamentary debate over the crisis
      in Abyssinia in 1935 that his lips were sealed on the matter, insinuating
      that he could not respond to criticisms without revealing state secrets: see
      Marguerite Potter, 'What Sealed Baldwin's Lips?', *Historian* 27, no. 1 (1964):
      21–36.
159   Security Service file related to Pitt-Rivers: KV 2/831. He had first attracted
      MI5 attention in 1934 when he invited Joyce to speak, though it was also
      reported that his anti-Semitic views had been known since 1930.

Chapter Six

1 See, for instance, letters with German correspondents in 1937/37: GPR 17/4.
2 Proceedings of divorce suit, Pitt-Rivers v. Pitt-Rivers, Sharpe intervening, 1937 (J 77/3773).
3 Letter from Wyndham Lewis to Pitt-Rivers, 9 May 1937 (GPR 18/2).
4 Proceedings of divorce suit, Pitt-Rivers v. Pitt-Rivers, Sharpe intervening, 1937 (J 77/3773).
5 See letters from Sharpe to Conwell-Evans, etc., 1937: GPR 25/2
6 Catherine Sharpe, 'Shadow over Czechoslovakia', *The Anglo-German Review* 1, no. 5 (1937): 202–203; George Pitt-Rivers, 'The Truth About Spain: A Remarkable Report', *The Anglo-German Review* 1, no. 4 (1937): 157–159.
7 Letter from Pitt-Rivers to Under Secretary of State for Foreign Affairs, 3 May 1937 (GPR 25/2).
8 *Hansard* House of Commons Debate 26 April 1937, vol 323 cc 6-12
9 'Centenary of the University of Göttingen', *Nature* 139, no. 3521 (1937): 701–703. Born took a position at Cambridge following his expulsion, turning down numerous other offers to do so. Correspondence relating to his expulsion can be found in Max Born Papers, Churchill Archives Centre, Cambridge, 1/2/1/9. He would win the Nobel Prize for Physics in 1954.
10 Pitt-Rivers reported the incident to the 18B panel considering his case: see transcript of 8 November 1940 hearing, pp. 15–16 (HO 45/25745). General Pitt-Rivers had received the honorary degree from Oxford, no doubt making his grandson desire the same qualification.
11 Draft letter from British delegation, authored by Pitt-Rivers, to Rector of the University, 29 June 1937 (GPR 50/2).
12 'Englische Adresse an Die Georgia Augusta', *Göttinger Nachrichten*, 30 June 1937. Clipping in GPR 50/2.
13 Ibid.
14 Minutes of IUSIPP Executive Committee, 27 July 1931 (Pitt-Rivers Private Papers).
15 IUSIPP reports on 1. Progress 2. Science of Population – Methodology and Classification (General Assembly, Paris, July 1937) (Pitt-Rivers Private Papers). For a complete treatment of the IUSIPP, see F. Landis MacKellar and Bradley W. Hart, 'Captain George Henry Lane-Fox Pitt-Rivers and the Prehistory of the IUSSP', *Population and Development Review* 40, no. 4 (2014): 653–675.
16 Minutes of IUSIPP General Assembly, 28 July 1931 (Pitt-Rivers Private Papers).
17 Letter from Pitt-Rivers to Oswald Mosley, 24 January 1940 (GPR 25/4).
18 Letter from Pitt-Rivers to Oberburgermeister of Frankfurt-am-Main, 14 August 1937 (GPR 17/4).
19 Unpublished autobiography of Catherine D. Taylor, 'From University to War' chapter, p. 46 (Private Collection).
20 Ibid., pp. 47–48 (Private Collection).
21 Ibid., p. 47 (Private Collection).
22 Catherine Sharpe, 'Hitler Girls in the New Germany', *The Anglo-German Review* II, no. 1 (1937): 17–18.

23 Richard Griffiths, *Fellow Travellers of the Right: British Enthusiasts for Nazi Germany: 1933–9* (London: Constable, 1980), 123, 224, 270.

24 Invitation of Sharpe and Pitt-Rivers to the 1937 Nuremberg Rally, 6 September 1937 (GPR 17/4).

25 Gerwin Strobl, *The Germanic Isle: Nazi Perceptions of Britain* (Cambridge: Cambridge University Press, 2000), 75–94.

26 Unpublished autobiography of Catherine D. Taylor, 'From University to War' chapter, pp. 42–43 (Private Collection).

27 Domvile Diaries, vol. 54, 10 September 1937. See also Evelyn Wrench, *Francis Yeats-Brown: 1886–1944* (London: Eyre & Spottiswoode, 1948), 218, 224.

28 Cf. Ibid., 224. Paragraphing dissolved.

29 Unpublished autobiography of Catherine D. Taylor, 'From University to War' chapter, p. 46 (Private Collection). For more on Unity Mitford's strange relationship with Hitler, see Jonathan Guinness and Catherine Guinness, *The House of Mitford* (London: Fontana, 1985); Mary S. Lovell, *The Mitford Girls: The Biography of an Extraordinary Family* (London: Little, Brown, 2001); David Rehak, *Hitler's English Girlfriend: The Story of Unity Mitford* (Stroud: Amberley, 2011).

30 A complete set of negatives of the photographs can be found in GPR 42.

31 Letter from Pitt-Rivers to Raymond Beazley, 28 September 1937 (GPR 25/2).

32 Ibid.

33 Letter from Catherine Sharpe to Raymond Beazley, 28 June 1938 (GPR 25/3). The delegate in question was identified as Major Watts.

34 Griffiths, 277–278.

35 Ibid., 275–277.

36 Letter from Raymond Beazley to Pitt-Rivers, 10 August 1937 (GPR 25/2).

37 Letter from Pitt-Rivers to Raymond Beazley, 28 September 1937 (GPR 25/2). See also Domvile Diaries, vol. 54, 9 September 1937: 'Pitt-Rivers has been trying to belittle the Link, I gather'.

38 Letter from Pitt-Rivers to Raymond Beazley, 28 September 1937 (GPR 25/2).

39 Griffiths, 277.

40 Sharpe, 'Hitler Girls in the New Germany'.

41 Unpublished autobiography of Catherine D. Taylor, 'From University to War' chapter, p. 43 (Private Collection).

42 Griffiths, 294–295.

43 Letter from GPR to A. Henry Higginson, 16 March 1938 (GPR 17/4).

44 Pitt-Rivers to Lothar Brökelmann, 16 March 1938 (GPR 17/4).

45 Pitt-Rivers to Maclaughlin, 16 March 1938 (GPR 17/4).

46 Letter from Pitt-Rivers to Adolf Hitler, 22 June 1938 (GPR 17/4).

47 Letter from H. Beblau, Legation Secretary, German Embassy, London, 22 July 1938 (GPR 17/4).

48 Letter from George Pitt-Rivers to Julian Pitt-Rivers, 30 December 1938 (Pitt-Rivers Private Papers).

49 George Pitt-Rivers, *The Story of the Ancient Manor of Hinton St Mary* (Farnham: The Pitt-Rivers Museum, 1947), 28.

50 Richard J. Evans, *The Third Reich in Power* (London: Penguin Books, 2005), 267.

51 Ibid., 671.

52   George Henry Lane Fox Pitt-Rivers, *The Czech Conspiracy: A Phase in the World-War Plot* (London: Boswell Pub. Co., 1938), 16.

53   Ibid., 18.

54   Ibid., 31, 33.

55   Ibid., Prefatory matter.

56   Evans, 672.

57   Ibid., 671–672.

58   Ibid. See also unpublished autobiography of Catherine D. Taylor, 'From University to War' chapter, p. 54 (Private Collection).

59   Cf. Ibid., 672.

60   Unpublished autobiography of Catherine D. Taylor, 'From University to War' chapter, pp. 52–53 (Private Collection).

61   Evans, 674.

62   Pitt-Rivers, *Czech Conspiracy*, 75, 87.

63   Ibid., 73–74.

64   Ibid., 87–88.

65   The Marquess of Londonderry was also a signatory: see ibid., 90. Londonderry made his own efforts to build Anglo-German understanding: see Ian Kershaw, *Making Friends with Hitler: Lord Londonderry, and Britain's Road to War* (London: Allen Lane, 2004).

66   See correspondence related to *The Czech Conspiracy* advance copies, 1938/39 (GPR 17/4).

67   Letters to Batault and Gario, December 1938 (GPR 17/4).

68   See letter acknowledging receipt from Maud Pearl, 11 March 1939 (GPR 13/2).

69   Letter from Pitt-Rivers to Neville Chamberlain, 18 March 1939 (GPR 13/2). Emphasis original.

70   Letter from Paymaster Rear Admiral W.E.R. Martin, 3 April 1939 (GPR 25/4).

71   Letter from Pelley Publishers to Pitt-Rivers, 6 March 1939 (GPR 13/2). See also Leo P. Ribuffo, *The Old Christian Right: The Protestant Far Right from the Great Depression to the Cold War* (Philadelphia: Temple University Press, 1983).

72   Letter from Gerald B. Winrod to Pitt-Rivers, 14 March 1939 (GPR 25/4).

73   Letter from Pitt-Rivers to Michael Pitt-Rivers, undated [1939] (GPR 27/5).

74   George Pitt-Rivers, *Your Home Is Threatened!* (Gillingham: T.H. Brickell & Son, 1939), 12.

75   Pitt-Rivers, *Your Home is Threatened*, 8.

76   Letter from Oscar Levy to George Pitt-Rivers, 12 September 1938 (GPR 19/2).

77   Undated MSS entitled 'Der Zukunft Europas' [The Future of Europe], undated (GPR 6/3).

78   Letter from Sharpe to Margaret Bothamley, 10 May 1938 (GPR 25/3).

79   Letter from Conwell-Evans to Sharpe, 23 June 1938 (GPR 25/3).

80   Letter from Sharpe to Margaret Bothamley, 10 May 1938 (GPR 25/3).

81   Letter from Beazley to Sharpe, 24 June 1938 (GPR 25/3).

82   Evans, 580–591.

83   Cf. Ibid., 591–592.

84   Letter from the AGF Council to members and the press, 18 November 1938 (GPR 25/3).

85   'German Treatment of Jews: Lord Mount Temple's Protest', *The Times*, 19 November 1938.

86   Letter from C.E. Carroll to Pitt-Rivers, 10 December 1938 (GPR 25/3).
87   Letter from Pitt-Rivers to Domvile, 29 January 1939 (GPR 25/4). Emphasis original.
88   See Carlton Jackson, *Who Will Take Our Children? The British Evacuation Program of World War Ii*, rev. ed. (Jefferson: McFarland, 2008).
89   Ibid., 18–20.
90   Ibid., 20–21.
91   Ibid., 22.
92   Ibid.
93   Indeed, the arrival of poor children created tension in some areas: ibid., 28–32.
94   Pitt-Rivers, *Your Home Is Threatened!*, 10.
95   Ibid., 11, 26.
96   Ibid., 12.
97   Ibid., 27.
98   Ibid.
99   'Squire won't have any refugees billeted: "Alien Habits in Rural Areas"' clipping, 22 February 1939 (GPR 3).
100   Clipping from *The Daily Sketch*, 23 February 1939 (GPR 3).
101   Clipping from Col. Blimp, *The Evening Standard*, 25 February 1939 (GPR 3).
102   'Statement of case against George Henry Lane Fox Pitt-Rivers', KV 2/831.
103   1 December 1937 MI5 minutes, noted in KV 2/831.
104   Unpublished autobiography of Catherine D. Taylor, 'From University to War' chapter, pp. 54–55 (Private Collection).
105   Letter from Sharpe (Ditchingham) to Mike Sharpe, undated [1939] (Private Collection).
106   Ibid
107   There is no evidence Michael made the journey, and the fact that he was already in the military by mid-1939 makes it unlikely that he could have obtained the leave to do so.
108   Unpublished autobiography of Catherine D. Taylor, 'From University to War' chapter, p. 55 (Private Collection).
109   A.W. Brian Simpson, *In the Highest Degree Odious: Detention without Trial in Wartime Britain* (Oxford: Clarendon Press, 1992), 216.
110   Transcript of 8 November 1940 hearing, p. 21 (HO 45/25745).
111   Letter from Chief Constable of Dorset to Home Office, 19 June 1940 (HO 45/25745).
112   Transcript of 11 November 1940 hearing, p. 19 (HO 45/25745).
113   Letter from Gardiner to Pitt-Rivers, 16 January 1939 (GPR 25/4).
114   Letter from Chief Constable of Dorset to Home Office, 19 June 1940 (HO 45/25745).
115   Copy of letter from Pitt-Rivers to A.T. Norris, 31 March 1940 (HO 45/25745); Report from J.G. Dickson to MI5, B.7 (KV 2/831).
116   Simpson, 209. Analysis of Fuller's political importance in Brian Holden Reid, *J.F.C. Fuller: Military Thinker*, Studies in Military and Strategic History (Basingstoke: Macmillan, 1987), 176.
117   Reid, 176.
118   Letter from Lord Alfred Douglas to Pitt-Rivers, 13 April 1939 (GPR 25/4). Emphasis original.

119   Evans, 691.
120   Ibid., 693.
121   Pitt-Rivers, *The Story of the Ancient Manor of Hinton St Mary*, 38.
122   Evans, 695.
123   Ibid., 699–701.
124   Ibid., 701–703.
125   Jackson, 35. See also Pitt-Rivers, *The Story of the Ancient Manor of Hinton St Mary*, 38.
126   Griffiths, 368.
127   Ibid., 372.
128   Richard C. Thurlow, *Fascism in Britain: A History, 1918–1985* (Oxford; New York: Blackwell, 1987), 199.
129   Rebecca West, *The Meaning of Treason* (London: Macmillan & Co. Ltd., 1949), 93, 98.
130   M.A. Doherty, *Nazi Wireless Propaganda: Lord Haw-Haw and British Public Opinion in the Second World War*, International Communications (Edinburgh: Edinburgh University Press, 2000), 12.
131   Thurlow, 191.
132   Ibid., 180–183.
133   Domvile Diaries, vol. 56
134   Ibid., 17 October 1939.
135   Home Office Advisory Committee Report, 21 November 1940 (HO 45/25745).
136   Dorset Constabulary Report, 19 June 1940 (HO 45/25745).
137   18 October 1939 MI5 minute 25a (KV 2/831).
138   Letter from A.T. Norris to Pitt-Rivers, March 1940 (HO 45/25745).
139   Letter from Pitt-Rivers to Norris, 31 March 1940 (HO 45/25745).
140   Letter from Pitt-Rivers to Mosley, 24 January 1940 (GPR 25/4).
141   Richard J. Evans, *The Third Reich at War* (London: Allen Lane, 2008), 118–120.
142   Ibid., 121–122.
143   Ibid., 123–133.
144   Griffiths, 354–355; Richard Griffiths, *Patriotism Perverted: Captain Ramsay, the Right Club, and British Anti-Semitism, 1939–40* (London: Constable, 1998).
145   Simpson, 146–157; Thurlow, 194–195.
146   Thurlow, 197–198; Simpson, 161–162.
147   Simpson, 70–80.
148   Cf. Ibid., 172.
149   Ibid., 173–188.
150   Letter from Chief Constable of Dorset to Home Office, 19 June 1940 (HO 45/25745).
151   Pitt-Rivers, *The Story of the Ancient Manor of Hinton St Mary*, 43–44.
152   Letter from A. Knight, Inspector, Dorset Constabulary, 6 July 1940 (KV 2/831).
153   'Reasons for order made under Defence Regulation 18B in the case of George Henry Lane Fox Pitt-Rivers' (KV 2/831).
154   Domvile Diaries, vol. 56, 27 June 1940.
155   Stephen Dorril, *Blackshirt: Sir Oswald Mosley and British Fascism* (London; New York: Penguin Books, 2007), 513. See also John Colville, *The Fringes*

of Power: Downing Street Diaries 1939–1955 (London: Hodder and Stoughton, 1985). The original list seems to be in HO 45/25747.

156 See the list in HO 45/25747.
157 Colville, 177. Interestingly, Colville's original diary seems to indicate that he first chose the adjective 'annoyed' to describe the Prime Minister's reaction to the appearance of these names on the detention list. It is also interesting that in the original he refers to Pitt-Rivers with the diminutive and familiar nickname 'Joe' rather than the usual 'Captain' or even 'George' (29 June 1940 entry, Jock Colville Diaries, Churchill Archives Centre 1/2).
158 J.G. Dickson to MI5, 2 July 1940 (KV 2/831).
159 Pitt-Rivers, *The Story of the Ancient Manor of Hinton St Mary*, 38.
160 Simpson, 245.
161 Letter from Pitt-Rivers to H.M. Principal Secretary of State for Home Affairs, 21 October 1940 (GPR 13/4).
162 Pamela Churchill, nee Digby, later Pamela Harriman, US ambassador to France.
163 *The Life and Death of King John,* Act IV, scene II. Pitt-Rivers later wrote that the quote 'seemed appropriate also to our democratic rulers of the present day' (Pitt-Rivers, *The Story of the Ancient Manor of Hinton St Mary*, 39).
164 Letter from Pitt-Rivers to Clementine Churchill, 18 September 1940 (GPR 13/4).
165 Letter from Levy to Pitt-Rivers, 1 July 1941 (GPR 18/2).
166 Ibid., 24 December 1940 (GPR 18/2).
167 Draft letter from Pitt-Rivers to Levy, 21 July 1940 (GPR 18/2).
168 Detainee John Charnley recalled the incident in his autobiographical account of those years: see John Charnley, *Blackshirts and Roses: An Autobiography* (London: Brockingday, 1990), 117.
169 MI5 objections to release from detention, 27 January 1941 (KV 2/831).
170 Pitt-Rivers, *The Story of the Ancient Manor of Hinton St Mary*, 39.

Chapter Seven

1 A.W. Brian Simpson, *In the Highest Degree Odious: Detention without Trial in Wartime Britain* (Oxford: Clarendon Press, 1992), 190.
2 George Pitt-Rivers, *The Story of the Ancient Manor of Hinton St Mary* (Farnham: The Pitt-Rivers Museum, 1947), 41.
3 Simpson, 190.
4 Ibid., 223.
5 MI5 objections to release from detention, 27 January 1941 (KV 2/831).
6 Letter from the Governor of Brixton Prison to the 18B Commissioners, 2 November 1940 (TS 27/514).
7 Simpson, 82–83.
8 Ibid., 83.
9 Claremont Haynes & Co. to the Under Secretary of State, Home Office, 29 July 1940 (GPR 13/6).
10 30 October 1940 Home Office minute (HO 45/25725).

11   Hearing transcript, 25 October 1940 (HO 45/25725).
12   Transcript of 8 November 1940 hearing, p. 9 (HO 45/25725).
13   Ibid., p. 12 (HO 45/25725).
14   Ibid., p. 13 (HO 45/25725).
15   Ibid., pp. 15–20 (HO 45/25725).
16   Ibid., p. 32 (HO 45/25725).
17   Ibid., p. 34 (HO 45/25725).
18   Ibid., p. 46 (HO 45/25725).
19   Transcript of 11 November 1940 hearing, p. 1 (HO 45/25725). Pitt-Rivers
     was incorrect to describe the Unknown Warrior as lying beneath the Cenotaph
     in Whitehall. The Warrior lies in Westminster Abbey, though the funeral
     procession involved the Cenotaph.
20   Transcript of 11 November 1940 hearing, p. 6 (HO 45/25725).
21   Ibid., pp. 11–12 (HO 45/25725).
22   Report of G.P. Churchill to the Home Office, 21 November 1940, point 12
     (HO 45/25725).
23   Ibid., points 10 and 11 (HO 45/25725).
24   Ibid., point 12 (HO 45/25725).
25   Ibid.
26   Minute of Home Office action, 26 November 1940 (HO 45/25725).
27   Ibid.
28   Minute of Home Office action, 5 January 1941 (HO 45/25725).
29   Ibid.
30   Report of G.P. Churchill to the Home Office, 23 January 1941
     (HO 45/25725).
31   Ibid.
32   MI5 objections to release from detention, 27 January 1941 (KV 2/831).
33   MI5 objections to release from detention, 27 January 1941, letter from John
     P.L. Redfern to the Home Office, 29 January 1941 (KV 2/831).
34   Home Office minute, 28 August 1941 (HO 45/25725).
35   Report of examination by W. Rowley Bristow, Orthopaedic Surgeon to St
     Thomas's, 22 September 1941 (GPR 13/5).
36   Letter from Carter Braine, 29 July 1941 (GPR 13/4).
37   30 August 1941 report of medical examination (HO 45/25725).
38   Letter from Pitt-Rivers to the Home Secretary, undated (GPR 18/1). The 'dot
     dot dot dash' was a reference to wartime BBC broadcasts that began with the
     opening notes of Beethoven's Fifth Symphony.
39   Simpson, 363–364.
40   The King v. Governor of Brixton Prison and another, Ex parte G.H.L.F. Pitt-
     Rivers, 19 December 1941 (TS 27/514).
41   Summary of case by Guy Liddell, 17 June 1940 (TS 27/514).
42   Copy of letter from Pitt-Rivers to Mosley, 11 June 1934 (TS 27/514).
43   The King v. Governor of Brixton Prison and another, Ex parte G.H.L.F. Pitt-
     Rivers, 19 December 1941, p. 12 (TS 27/514).
44   Ibid.
45   Richard C. Thurlow, *Fascism in Britain: A History, 1918–1985* (Oxford, UK;
     New York: Blackwell, 1987), 213–214.
46   Ibid., 230–231.

47 Telegram from Churchill to Clement Attlee and Herbert Morrison, 25 November 1943 (Chartwell Papers 20/130/16-17).

48 See description of press cuttings related to Pitt-Rivers case, 1941–42 (GPR 25/5).

49 Letter from Pitt-Rivers to Morier, 25 January 1942; letter from R.B.T. Keen to Pitt-Rivers, 2 March 1942 (GPR 13/5).

50 Warrant and order from MI5 to the Postmaster General, 19 February 1942 (KV 2/831).

51 Copy of letter from Keith to Pitt-Rivers, 9 February 1942 (GPR 13/3).

52 Copy of letter from Keith to Pitt-Rivers, 17 February 1942 (GPR 13/3).

53 Arthur Keith, *An Autobiography* (London: Watts, 1950), 553. Interestingly, Keith also stated that Pitt-Rivers 'joined the party of Sir Oswald Mosley' in the same footnote. Pitt-Rivers marked this passage in his personal copy of Keith's autobiography and wrote 'error!' in the margin. Even years after his detention had ended, Pitt-Rivers chafed at the claim that he had ever been part of the BUF.

54 Letter from Levy to Pitt-Rivers, 2 February 1942 (GPR 13/5).

55 Letter from Levy to Pitt-Rivers, 1 September 1941 (GPR 18/2).

56 Ibid.

57 Letter from Levy to Pitt-Rivers, 6 June 1942 (GPR 13/5).

58 Letter from Birkett to Home Office, 2 March 1942 (HO 283/31).

59 Letter from Keen to Pitt-Rivers, 23 March 1942 (KV 2/831).

60 Supplementary report on the case of George Henry Lane Fox Pitt-Rivers, 23 April 1942 (HO 283/58). This file remained closed until opened by a Freedom of Information Act request in the course of this research. Various redactions remain in the file.

61 Testimony of Mrs Astley-Corbett before the 18B Advisory Committee, 16 April 1942 (HO 283/58).

62 Supplementary report on the case of George Henry Lane Fox Pitt-Rivers, 23 April 1942 (HO 283/58).

63 Ibid.

64 Malcolm Muggeridge, *Chronicles of Wasted Time: The Infernal Grove*, vol. 2 (London: Fontana, 1975), 117.

65 See letters related to billeting of manor house, 1943–45 (GPR 17/2).

66 Letter from Pitt-Rivers to Director of Quartering, 17 November 1943 (GPR 17/1).

67 Ibid.

68 Letter from Pitt-Rivers to Thynne, 2 March 1942 (GPR 23/3).

69 Letter from Thynne to Pitt-Rivers, 19 March 1942 (GPR 23/3).

70 See letter from Pitt-Rivers to Hippesley-Cox offering data on South Pacific to intelligence division (1942) – report on Aua Island (GPR 18/3).

71 Letter from Firth to Pitt-Rivers, 2 September 1942 (GPR 18/3).

72 Pitt-Rivers, 44.

73 Ibid. See also letter from Pitt-Rivers to Lady Anderson, 5 May 1944 (GPR 27/5) and letter from Pitt-Rivers to Sir John Anderson, 29 June 1945 (GPR 13/3).

74 Letter from Pitt-Rivers to Joe Beckett, Jr., 28 November 1945 (GPR 13/3).

75 'The S.R. case' summary document, KV 2/3800. Roberts' identity was only released to the public in late 2014: see KV 2/3874.

76   'The S.R. case' summary document, KV 2/3800.
77   Pitt-Rivers, 45.
78   Letter from Pitt-Rivers to Julian Pitt-Rivers, 12 July 1948 (Pitt-Rivers Private Papers).
79   See, for instance, GPR 19/5 for 1945 correspondence with *The Times*
80   The Pitt-Rivers connection was reported to Sir Thomas Vansittart in December 1945 by informants (Vansittart Papers, Set II, 1/24). Reference to the 'Hebrew economy' found in GPR 19/4.
81   Pitt-Rivers, 45.
82   Matthew Sweet, *The West End Front: The Wartime Secrets of London's Grand Hotels* (London: Faber, 2012), 217.
83   Ibid., 225–226.
84   Ibid., 227–228.
85   Ibid., 228.
86   Ibid., 230.
87   Home Office 18B Advisory Committee report, 6 December 1944 (HO 45/25745).
88   Sweet, 233–234.
89   Ibid., 235–237.
90   Home Office 18B Advisory Committee report, 6 December 1944 (HO 45/25745).
91   Sweet, 240.
92   Ibid., 242–244.
93   Ibid., 247.
94   Home Office 18B Advisory Committee report, 6 December 1944 (HO 45/25745).
95   Julian Alfred Pitt-Rivers, *The Story of the Royal Dragoons, 1938–1945* (London: Published for the Royal Dragoons by W. Clowes, 1945).
96   Pitt-Rivers, 46.
97   See Foreign Office reports related to Faisal: FO 371/61679, 371/61684
98   'Obituary: Lady Rumbold: Hermione Baddeley's Daughter and Society Beauty Who Enchanted the Young Lucian Freud', *The Telegraph*, http://www.telegraph.co.uk/news/obituaries/3815125/Lady-Rumbold.html (accessed 9 October 2014).
99   Letter from Pitt-Rivers to Julian Pitt-Rivers, 11 July 1946 (Pitt-Rivers Private Papers).
100   Letter from Pitt-Rivers to Julian Pitt-Rivers, 12 July 1948 (Pitt-Rivers Private Papers).
101   Sweet, 251.
102   Letter from Pitt-Rivers to Pauline Pitt-Rivers, 8 February 1949 (Pitt-Rivers Private Papers).
103   Julian Alfred Pitt-Rivers, *The People of the Sierra* (London; Beccles: William Clowes and Sons, Ltd., 1954).
104   E.E. Evans-Pritchard, 'Foreword', in *The People of the Sierra*, ed. Julian Alfred Pitt-Rivers (London; Beccles: William Clowes and Sons, Ltd., 1954), ix. See also John Corbin, 'Obituary: Julian Pitt-Rivers: Scholar Who Opened a New Chapter in Social Anthropology', *The Guardian*, http://www.theguardian.com/news/2001/sep/14/guardianobituaries.socialsciences (accessed 10 October 2014).

105 'Lord Montagu on the Court Case Which Ended the Legal Persecution of Homosexuals', *Mail Online*, http://www.dailymail.co.uk/news/article-468385/Lord-Montagu-court-case-ended-legal-persecution-homosexuals.html (accessed 14 December 2009).

106 'From our special correspondent', 'Lord Montagu in the Box.', *The Times*, 20 March 1954, 3; 'Charges against Three Men', *The Times*, 26 January 1954, 3.

107 'Lord Montagu on the Court Case Which Ended the Legal Persecution of Homosexuals'.

108 Ian Gilmour, 'The Montagu Case', *The Spectator*, 25 November 1955, 10–13.

109 'From our special correspondent', 'Three Men Sent to Prison' *The Times*, 25 March 1954, 4.

110 Committee on Homosexual Offences and Prostitution, *Report of the Committee on Homosexual Offences and Prostitution* (London: Her Majesty's Stationery Office, 1957); Peter Wildeblood, *Against the Law* (London: Weidenfeld and Nicolson, 1956).

111 See Hilary Spurling, *The Girl from the Fiction Department: A Portrait of Sonia Orwell* (London; New York: Hamish Hamilton, 2002).

112 Michael Pitt Rivers, *Dorset. A Shell Guide*, ed. Paul Nash, New ed. (London: Faber, 1966).

113 Sweet, 252. Also sourced from private information.

114 J.M.C. Toynbee, 'A New Roman Mosaic Pavement Found in Dorset', *The Journal of Roman Studies* 54, no. 1–2 (1964).

115 George Henry Lane Fox Pitt-Rivers, *The Riddle of the 'Labarum' and the Origin of Christian Symbols* (London: Allen & Unwin, 1966).

116 Henry Chadwick, 'George Pitt-Rivers: The Riddle of the 'Labarum' and the Origin of Christian Symbols. p. 92. London: Allen & Unwin, 1966. Cloth, 35s.', *The Classical Review (New Series)* 17, no. 2 (1967): 234.

117 Sweet, 253.

118 Anonymous, 'Capt. G.H. Pitt-Rivers (Obituary)', *The Times*, 18 June 1966, 12.

119 'In Memorium', *The Times*, 17 June 1981, 26.

Afterword

1 'A Staff Reporter', 'Pitt-Rivers Collection for Museum', *The Times*, 24 January 1975, 3.

2 See, for instance, Michael Pitt-Rivers, 'Cultural General', *Books and Bookmen* 22, no. 9 (1977): 23–35.

3 Philip Hoare, 'Obituary: Michael Pitt-Rivers', *The Independent*, 11 January 2000.

4 John Corbin, 'Obituary: Julian Pitt-Rivers: Scholar Who Opened a New Chapter in Social Anthropology', *The Guardian*, http://www.theguardian.com/news/2001/sep/14/guardianobituaries.socialsciences (accessed 10 October 2014).

5 For a copy of her co-authored report on the Belsen inmates, see Wellcome Library, Dame Janet Vaughan Papers, GC 186/7.

6   J.R. Tata, 'Rosalind Venetia Pitt-Rivers', The Royal Society, http://thyroid.
     org/wp-content/uploads/2012/04/Rosalind_Venetia_Pitt-Rivers.pdf
     (accessed 13 June 2014); Jamshed R. Tata, 'Rivers, Rosalind Venetia Lane Fox
     Pitt- (1907–1990)', *Oxford Dictionary of National Biography*, http://www.
     oxforddnb.com/view/article/57570 (accessed 13 June 2014).
7   Matthew Sweet, *The West End Front: The Wartime Secrets of London's Grand
     Hotels* (London: Faber, 2012), 253. Also private information.
8   'The George Pitt-Rivers Laboratory for Bio-Archaeology', University of
     Cambridge, http://www.arch.cam.ac.uk/research/laboratories/pitt-rivers
     (accessed 11 October 2014).
9   Catherine Taylor, *If Courage Goes: My Twenty Years in South African Politics*
     (Johannesburg: Macmillan, 1976), 35–42.
10   Guy Liddell Diaries, 23 May 1941, p. 913 (KV 4/187).
11   Taylor, 13–14.
12   Ibid., 32–33.
13   Lesley A. Hall, 'Racial Science and British Society, 1930–62', *Social History
     of Medicine* 22, no. 2 (2009): 410–411; Gavin Schaffer, '"Like a Baby with a
     Box of Matches": British Scientists and the Concept of "Race" in the Inter-
     War Period', *The British Journal for the History of Science* 38, no. 3 (2005):
     307–324; Gavin Schaffer, '"Scientific Racism Again?": Reginald Gates, the
     Mankind Quarterly and the Question of "Race" in Science after the Second
     World War', *Journal of American Studies* 41, no. 2 (2007): 253; Gavin
     Schaffer, *Racial Science and British Society, 1930–62* (Basingstoke: Palgrave
     Macmillan, 2008).
14   Schaffer, 253.
15   Pitt-Rivers, 'Cultural General', 23.

BIBLIOGRAPHY

Archival Collections

Archiv der Max-Planck-Gesellschaft (MPG-Archiv), Berlin
Papers of Central Institutions of the Kaiser Wilhelm Society: Administrative
    Headquarters (Abteilung I, Repositorium IA)
Papers of the Kaiser Wilhelm Institute for Anthropology, Human Genetics and
    Eugenics (Ab. I, Rep. 3)
Baker Library, Harvard University
Papers of Elton Mayo
Bancroft Library, University of California, Berkeley
Robert H. Lowie Papers
*The British Library of Political and Economic Science, London School of
    Economics*
Papers of Raymond Firth
Papers of Bronislaw Malinowski
Churchill Archives Centre, Cambridge
Papers of Max Born
Papers of Randolph Churchill
Chartwell Papers of Sir Winston Spencer Churchill
Diaries of John 'Jock' Colville
Papers of George Henry Lane Fox Pitt-Rivers
Papers of Sir Robert Vansittart
King's College London Archive
Papers of Reginald Ruggles Gates
National Archives of Australia
Department of Home Affairs Papers (series A1)
National Archives of the United Kingdom
Foreign Office (FO)
Home Office (HO)
Supreme Court of Judicature (J)
Security Service (KV)
Prime Minister's Office (PREM)
Treasury Solicitor and HM Procurator General's Department (TS)
War Office Papers (WO)
National Maritime Museum Archives, Greenwich, London
Diaries of Admiral Sir Barry Domvile
Pickler Memorial Library, Truman State University
Papers of Harry H. Laughlin

Private Collections
Personal papers of George Henry Lane Fox Pitt-Rivers
Personal papers and unpublished manuscript chapters of Catherine Dorothea
    Taylor (nee Sharpe)
The Royal College of Surgeons, London
Papers of Sir Arthur Keith
University of California, Los Angeles (UCLA) Library Special Collections
Papers of Ferdinand C.S. Schiller (collection 191)
University Library Manuscripts Room, Cambridge University
Papers of the Darwin family
The Wellcome Library, London
Papers of Carlos Paton (C.P.) Blacker
Papers of Dame Janet Vaughan
Papers of the Eugenics Education Society/Eugenics Society/Galton Institute
Papers of the Population Investigation Committee (PIC)
Woodson Research Center, Fondren Library, Rice University, Texas
Papers of Julian Sorell Huxley (HUX)
Yale University Library
Papers of Bronislaw Malinowski

Published Works

Allen, Garland E. 'The Eugenics Record Office at Cold Spring Harbor, 1910-1940:
    An Essay in Institutional History'. *Osiris* 2 (1986): 225–264.
Anonymous. 'Mental Deficients'. *The Birth Control News* VIII, no. 1 (1929): 2.
Anonymous. 'Sterilization Laws: More Than One-Half of the States Have
    Eugenical Sterilization Laws'. *Eugenical News* XVI, no. 8 (1931): 129.
Anonymous. 'Eugenics and Religion: Summary of a Debate Arranged by the
    Eugenics Society (April 25th, 1933)'. *The Eugenics Review* XXV, (1933–1934):
    101–103.
Anonymous. 'A "Fourth Party" in Dorset: Sir C. Hanbury's Varied Opponents'. *The
    Times*, 31 October 1935, 9, col. G.
Anonymous. 'North Dorset by-Election: The Prime Minister's Message'. *The Times*,
    9 July 1937, 9, col. D.
Anonymous. 'Capt. G.H. Pitt-Rivers (Obituary)'. *The Times*, 18 June 1966, 12,
    col. G.
Anonymous. 'Obituary: Carlos Paton Blacker'. *The Lancet* 305, no. 7915 (1975):
    1096–1097.
Asquith, H. H., M. G. Brock and Eleanor Brock. *Letters to Venetia Stanley*.
    Oxford: Oxford University Press, 1985.
Atkinson, C. T. *History of the Royal Dragoons, 1661–1934*. Glasgow: Printed for
    the Regiment by R. Maclehose and Co Ltd. at the University Press, 1934.
'Attacks on Tithe System'. *The Manchester Guardian*, 11 January 1935, 7.
Barkan, Elazar. *The Retreat of Scientific Racism: Changing Concepts of Race in
    Britain and the United States between the World Wars*. Cambridge: Cambridge
    University Press, 1992.
Bashford, Alison and Philippa Levine. *The Oxford Handbook of the History of
    Eugenics*. New York, Oxford: Oxford University Press, 2010.

Black, Edwin. *War Against the Weak: Eugenics and America's Campaign to Create a Master Race*. New York, London: Four Walls Eight Windows, 2003.

Blacker, C. P. *Birth Control and the State; a Plea and a Forecast* To-Day and To-Morrow. London: K. Paul, Trench, 1926.

Blacker, C. P. *Have You Forgotten Yet? The First World War Memoirs of C.P. Blacker*, edited by J. G. C. Blacker. London: Leo Cooper, 2000.

Blake, Robert, 'Churchill, Randolph Frederick Edward Spencer (1911–1968)'. *Oxford Dictionary of National Biography*. http://www.oxforddnb.com/view/article/37283 (accessed 12 July 2014).

Bland, Lucy. 'British Eugenics and "Race Crossing": A Study of an Interwar Investigation'. *New Formations* 60, (2006): 66–78.

Bowden, Mark. *Pitt Rivers: The Life and Archaeological Work of Lieutenant-General Augustus Henry Lane Fox Pitt Rivers, DCL, FRS, FSA*. Cambridge, UK, New York: Cambridge University Press, 1991.

Brace, Laura. *The Idea of Property in Seventeenth-Century England: Tithes and the Individual*. Politics, Culture and Society in Early Modern Britain. Manchester: Manchester University Press, 1998.

Briant, Keith. *Marie Stopes, a Biography*. London: Hogarth Press, 1962.

Brock, Michael, 'Stanley, (Beatrice) Venetia (1887–1948)'. *Oxford Dictionary of National Biography*. http://www.oxforddnb.com/view/article/41069 (accessed 13 June 2014).

Brody, Leslie. *Irrepressible: The Life and Times of Jessica Mitford*. 1 vol. Washington, London: Counterpoint, 2011.

Brose, Eric Dorn. *A History of the Great War: World War One and the International Crisis of the Early Twentieth Century*. New York: Oxford University Press, 2010.

Brown, W. J. *The Communist Movement and Australia: An Historical Outline, 1890s to 1980s*. Haymarket: Australian Labor Movement History Publications, 1986.

Burleigh, R. and J. Clutton-Brock. 'Pitt Rivers and Petrie in Egypt'. *Antiquity* 56, (1982): 208–208.

Buschmann, Rainer F. *Anthropology's Global Histories: The Ethnographic Frontier in German New Guinea, 1870–1935*. Perspectives on the Global Past. Honolulu: University of Hawai'i Press, 2009.

Cabada, Ladislav. 'From Munich to the Renewal of Czechoslovakia'. In *Czechoslovakia and the Czech Republic in World Politics*, edited by Ladislav Cabada and Sárka Waisová, 33–50. Lanham: Lexington Books, 2011.

Calder-Marshall, Arthur. *Havelock Ellis: A Biography*. London: R. Hart-Davis, 1959.

Carr, Richard. 'The Phoenix Generation at Westminster: Great War Veterans Turned Tory Mps, Democratic Political Culture, and the Path of British Conservatism from the Armistice to the Welfare State'. Unpublished PhD thesis, University of East Anglia, 2010.

Carr, Richard. *Veteran MPs and Conservative Politics in the Aftermath of the Great War: The Memory of All That*. Farnham, Burlington: Ashgate Pub., 2013.

Carr, Richard and Bradley W. Hart. 'Old Etonians, Great War Demographics and the Interpretations of British Eugenics, C.1914–1939'. *First World War Studies* 3, no. 2 (2012): 217–239.

'Centenary of the University of Göttingen'. *Nature* 139, no. 3521 (1937): 701–703.

Chadwick, Henry. 'George Pitt-Rivers: The Riddle of the "Labarum" and the Origin of Christian Symbols. Pp. 92. London: Allen & Unwin, 1966. Cloth, 35s'. *The Classical Review (New Series)* 17, no. 2 (1967): 234.

Challis, Debbie. *The Archaeology of Race: The Eugenic Ideas of Flinders Petrie and Francis Galton*. London, New York: Bloomsbury Academic, 2013.

'Charges against Three Men'. *The Times*, 26 January 1954, 3.

Charnley, John. *Blackshirts and Roses: An Autobiography*. London: Brockingday, 1990.

Chesterton, A. K. *Oswald Mosley, Portrait of a Leader*. London: Action Press, Ltd., 1937.

Chesterton, G. K. *Eugenics and Other Evils*. London, New York: Cassell, 1922.

Churchill, Winston S. *His Father's Son: The Life of Randolph Churchill*. London: Weidenfeld & Nicolson, 1996.

Clark, C. M. H. *A History of Australia: Vol. 6*. Melbourne: Melbourne University Press, 1987.

Clark, Christopher M. *The Sleepwalkers: How Europe Went to War in 1914*. First U.S. ed. New York: Harper, 2013.

Cohn, Norman. *Warrant for Genocide: The Myth of the Jewish World-Conspiracy and the Protocols of the Elders of Zion*. London: Eyre & Spottiswoode, 1967.

Collis, John Stewart. *An Artist of Life: A Study of the Life and Work of Havelock Ellis*. London: Cassell, 1959.

Colville, John. *The Fringes of Power: Downing Street Diaries 1939–1955*. London: Hodder and Stoughton, 1985.

Comfort, Nathaniel C., 'Jan A. Witkowski and John R. Inglis (Editors). Davenport's Dream: Twenty-First Century Reflections on Heredity and Eugenics'. *Isis* 100, no. 1 (2009): 191–192.

Comfort, Nathaniel C. *The Science of Human Perfection: How Genes Became the Heart of American Medicine*. New Haven: Yale University Press, 2012.

Committee on Homosexual Offences and Prostitution. *Report of the Committee on Homosexual Offences and Prostitution*. London: Her Majesty's Stationery Office, 1957.

Cooper, John W. 'The Preservation of Native Races'. *Journal of Heredity* 19, no. 5 (1928): 234.

Corbin, John, 'Obituary: Julian Pitt-Rivers: Scholar Who Opened a New Chapter in Social Anthropology', *The Guardian*, http://www.theguardian.com/news/2001/sep/14/guardianobituaries.socialsciences (accessed 10 October 2014).

Cosway, A. H. *Guide to the Tithe Act, 1936*. London: Sir I. Pitman & Sons, Ltd., 1937.

Coupland, Philip. 'H.G. Wells's "Liberal Fascism"'. *Journal of Contemporary History* 35, no. 4 (2000): 541–558.

Crozier, Ivan. 'Havelock Ellis, Eugenicist'. *Studies in History and Philosophy of Science Part C: Studies in History and Philosophy of Biological and Biomedical Sciences* 39, no. 2 (2008): 187–194.

Cunneen, Chris, 'Forster, Sir Henry William (1866–1936)', National Centre of Biography, Australian National University, http://adb.anu.edu.au/biography/forster-sir-henry-william-6213/text10681 (accessed 3 June 2014).

Dalton, F. T. 'New Books & Reprints: Political'. *The Times Literary Supplement*, no. 976 (1920): 638.

Davis, Peter. 'How All Souls Got Its Anthropologist'. In *A History of Oxford Anthropology*, edited by Peter Rivière, 62–76. New York: Berghahn Books, 2007.

Doherty, M. A. *Nazi Wireless Propaganda: Lord Haw-Haw and British Public Opinion in the Second World War* International Communications. Edinburgh: Edinburgh University Press, 2000.

Domvile, Barry. *By and Large*. London: Hutchinson, 1936.

Domvile, Barry. *From Admiral to Cabin Boy*. London: Boswell Pub. Co., 1947.

Donaldson, Frances Lonsdale. *Edward VIII*. London: Weidenfeld & Nicolson, 1974.

Dorril, Stephen. *Blackshirt: Sir Oswald Mosley and British Fascism*. London, New York: Penguin Books, 2007.

'Englische Adresse an Die Georgia Augusta'. *Göttinger Nachrichten*, 30 June 1937, 1.

Evans, Richard J. *The Coming of the Third Reich*. London: Penguin Books, 2003.

Evans, Richard J. *The Third Reich in Power*. London: Penguin Books, 2005.

Evans, Richard J. *The Third Reich at War*. London: Allen Lane, 2008.

Evans-Pritchard, E. E. 'Foreword'. In *The People of the Sierra*, edited by Julian Alfred Pitt-Rivers, 232 p. London, Beccles: William Clowes and Sons, Ltd., 1954.

Firth, Raymond. *Primitive Economics of the New Zealand Maori*. London: G. Routledge, 1929.

FitzRoy, Almeric William. *Memoirs*. Vol. 1. 2 vols. 6th ed. London: Hutchinson, 1925.

FitzRoy, Almeric William. *Memoirs*. Vol. 2. 2 vols. 6th ed. London: Hutchinson, 1925.

Freeman-Ishill, Rose. *Havelock Ellis*. Berkeley Heights: Oriole Press, 1942.

Freud, Sigmund and A. A. Brill. *Totem and Taboo: Resemblances between the Psychic Lives of Savages and Neurotics*. London: G. Routledge & Sons, 1919.

Freud, Sigmund and Joan Riviere. *Civilization and Its Discontents*. The International Psycho-Analytical Library. London: Hogarth Press, Institute of Psycho-Analysis, 1930.

Galton, Francis. *Hereditary Genius: An Inquiry into Its Laws and Consequences*. London: Macmillan, 1869.

Galton, Francis. *Essays in Eugenics*. London: Eugenics Education Society, 1909.

Gardiner, Juliet. *The Thirties: An Intimate History of Britain*. London: HarperPress, 2011.

'The George Pitt-Rivers Laboratory for Bio-Archaeology'. University of Cambridge. http://www.arch.cam.ac.uk/research/laboratories/pitt-rivers (accessed 11 October 2014).

'German Treatment of Jews: Lord Mount Temple's Protest'. *The Times*, 19 November 1938, 7.

'Germans and Czechs'. *The Times*, 29 August 1936, 9.

Gilman, Sander L. 'Fat as Disability: The Case of the Jews'. *Literature and Medicine* 23, no. 1 (2004): 46–60.

Gilmour, Ian. 'The Montagu Case'. *The Spectator*, 25 November 1955, 10–13.

Gosney, Ezra Seymour and Paul Bowman Popenoe. *Sterilization for Human Betterment; a Summary of Results of 6,000 Operations in California, 1909–1929*. New York: The Macmillan company, 1931.

Gregory, Adrian. *The Last Great War: British Society and the First World War.* Cambridge, New York: Cambridge University Press, 2008.

Griffiths, Richard. *Fellow Travellers of the Right: British Enthusiasts for Nazi Germany: 1933–9.* London: Constable, 1980.

Griffiths, Richard. *Patriotism Perverted: Captain Ramsay, the Right Club, and British Anti-Semitism, 1939–40.* London: Constable, 1998.

Guinness, Jonathan and Catherine Guinness. *The House of Mitford.* London: Fontana, 1985.

Hall, Lesley A. 'Racial Science and British Society, 1930–62'. *Social History of Medicine* 22, no. 2 (2009): 410–411.

Hall, Ruth E. *Marie Stopes: A Biography.* London: Deutsch, 1977.

Hall, Ruth E. *Dear Dr Stopes: Sex in the 1920s.* Harmondsworth: Penguin, 1981.

Harmsen, Hans and Franz Lohse. *Bevölkerungsfragen: Bericht Des Internationalen Kongresses Für Bevölkerungswissenschaft, Berlin, 26. August-1, September 1935.* München: J. F. Lehmann, 1936.

Hart, Bradley W. 'Science, Politics, and Prejudice: The Dynamics and Significance of British Anthropology's Failure to Confront Nazi Racial Ideology'. *European History Quarterly* 43, no. 2 (2013): 301–325.

Hastings, Selina. *Nancy Mitford: A Biography.* Vintage Lives. London: Vintage, 2002.

Hawkins, Desmond. *Concerning Agnes: Thomas Hardy's 'Good Little Pupil'.* Gloucester: A. Sutton, 1982.

Hoare, Philip. 'Obituary: Michael Pitt-Rivers'. *The Independent*, 11 January 2000, 6.

Hodson, Cora B. S. *Human Sterilization to-Day: A Survey of the Present Position.* London: Watts & co., 1934.

Holmes, S. J. 'In Defense of Eugenics'. *Journal of Heredity* 23, no. 6 (1932): 243–244.

'In Memorium'. *The Times*, 17 June 1981, 26.

'Interim Report of the Proceedings of the First General Assembly of the International Union for the Scientific Investigation of Population Problems'. *Journal of the American Statistical Association* 23, no. 163 (1928): 306–317.

Ishiguro, Kazuo. *The Remains of the Day.* 1st American ed. New York: Knopf: Distributed by Random House, 1989.

Jackson, Carlton. *Who Will Take Our Children? The British Evacuation Program of World War Ii.* Rev. ed. Jefferson: McFarland, 2008.

Jarvie, Ian Charles. *Functionalism* Basic Concepts in Anthropology. Minneapolis: Burgess Publishing Company, 1973.

Jefferies, Matthew and Mike Tyldesley. *Rolf Gardiner: Folk, Nature and Culture in Interwar Britain.* Farnham: Ashgate, 2011.

Jouvenal, Bertrand de. 'Heart to Heart Talk with Hitler'. *The Anglo-German Review* 1, no. 4 (1937): 1.

Keith, Arthur. 'Introduction'. In *Weeds in the Garden of Marriage*, edited by George Henry Lane Fox Pitt-Rivers, ix–xiii. London: N. Douglas, 1931.

Keith, Arthur. *The Place of Prejudice in Modern Civilization (Prejudice and Politics): Being the Substance of a Rectorial Address to the Students of Aberdeen University.* London: Williams & Norgate Ltd., 1931.

Keith, Arthur. *An Autobiography.* London: Watts, 1950.

Kershaw, Ian. *Making Friends with Hitler: Lord Londonderry, and Britain's Road to War.* London: Allen Lane, 2004.

Kevles, Daniel J. *In the Name of Eugenics: Genetics and the Uses of Human Heredity*. Cambridge, MA: Harvard University Press, 1995.

Kühl, Stefan. *The Nazi Connection: Eugenics, American Racism, and German National Socialism*. New York, Oxford: Oxford University Press, 1994.

Kühl, Stefan. *Die Internationale Der Rassisten: Aufstieg Und Niedergang Der Internationalen Bewegung Für Eugenik Und Rassenhygiene Im 20. Jahrhundert*. Frankfurt: Campus Verlag, 1997.

Kühl, Stefan. 'The Relationship between Eugenics and the So-Called "Euthanasia Action" in Nazi Germany: A Eugenically Motivated Peace Policy and the Killing of the Mentally Handicapped During the Second World War'. In *Science in the Third Reich*, edited by Margit Szöllösi-Janze, 185–210. Oxford: Berg, 2001.

Kühl, Stefan. *For the Betterment of the Race: The Rise and Fall of the International Movement for Eugenics and Racial Hygiene*. New York, NY: Palgrave Macmillan, 2013.

'Letters to the Editor'. *The Anglo-German Review* 1, no. 2 (1936): 96.

Levine, Naomi B. *Politics, Religion, and Love: The Story of H.H. Asquith, Venetia Stanley and Edwin Montagu*. New York, London: New York University Press, 1991.

Levy, Oscar. *The Revival of Aristocracy*. Translated by Leonard A. Magnus. London: Probsthain & Co., 1906.

Levy, Oscar. 'The Nietzsche Movement in England: A Retrospect, a Confession, and a Prospect'. *The New Age* XII, no. 7 (1912): 157–158.

Levy, Oscar. 'Nietzsche and the Jews II'. *The New Age* XVI, no. 8 (1914): 193–195.

Levy, Oscar. 'Prefatory Letter'. In *The World Significance of the Russian Revolution*, edited by George Pitt-Rivers, i–xiii. Oxford: Basil Blackwell, 1920.

Levy, Oscar. 'Mussolini Says: "Live Dangerously": Italian Premier and Leader of Il Fascismo Acknowledges Nietzsche as Spritual Master and Avows That He Has Tried to Follow His Precept'. *The New York Times*, 9 November 1924, SM7.

Levy, Oscar, Leila Kais, Steffen Dietzsch and Julia V. Rosenthal. *Gesammelte Schriften und Briefe*. 1. Aufl. ed. Berlin: Parerga, 2005.

Lidbetter, Ernest James. *Heredity and the Social Problem Group*. London: E. Arnold, 1933.

Lombardo, Paul A. *Three Generations, No Imbeciles: Eugenics, the Supreme Court, and Buck V. Bell*. Baltimore, London: Johns Hopkins University Press, 2010.

'Lord Montagu in the Box'. *The Times*, 20 March 1954, 3.

'Lord Montagu on the Court Case Which Ended the Legal Persecution of Homosexuals'. *Mail Online*. http://www.dailymail.co.uk/news/article-468385/Lord-Montagu-court-case-ended-legal-persecution-homosexuals.html (accessed 14 December 2009).

Love, Gary. '"What's the Big Idea?": Oswald Mosley, the British Union of Fascists and Generic Fascism'. *Journal of Contemporary History* 42, no. 3 (2007): 447–468.

Lovell, Mary S. *The Mitford Girls: The Biography of an Extraordinary Family*. London: Little, Brown, 2001.

Lowie, Robert Harry. *Are We Civilized?: Human Culture in Perspective*. New York: Harcourt, Brace and Company, 1929.

Ludovici, Anthony M. 'On 'Conscience and Fanaticism''. *The New Age* 25, no. 24 (1919): 395–396.

MacKellar, F. Landis and Bradley W. Hart. 'Captain George Henry Lane-Fox Pitt-Rivers and the Prehistory of the IUSSP'. *Population and Development Review* 40, no. 4 (2014): 653–675.

Macnicol, John. 'Eugenics and the Campaign for Voluntary Sterilization in Britain between the Wars'. *Social History of Medicine* 2, no. 2 (1989): 147–169.

Mairet, Philip. *A. R. Orage: A Memoir*. London: J.M. Dent and Sons, Ltd., 1936.

Malinowski, Bronislaw. *Argonauts of the Western Pacific: An Account of Native Enterprise and Adventure in the Archipelagoes of Melanesian New Guinea*. London, New York: George Routledge & Sons, Ltd., 1922.

Malinowski, Bronislaw, Elsie Masson and Helena Wayne. *The Story of a Marriage: The Letters of Bronislaw Malinowski and Elsie Masson*. Vol. 2. 2 vols. London: Routledge, 1995.

Malthus, T. R. *An Essay on the Principle of Population, as It Affects the Future Improvement of Society. With Remarks on the Speculations of Mr. Godwin, M. Condorcet, and Other Writers*. London: Printed for J. Johnson, 1798.

'Mary Hinton (I) (1896–1979)', IMDB. http://www.imdb.com/name/nm0386003/ (accessed 8 June 2014).

Maude, Aylmer. *The Authorized Life of Marie C. Stopes*. London: Williams & Norgate Ltd., 1924.

Mencken, H. L. *The Philosophy of Friedrich Nietzsche*. London: T.F. Unwin, 1908.

Millar, Ronald. *The Piltdown Mystery: The Story Behind the World's Greatest Archaeological Hoax*. Seaford: S.B., 1998.

Millar, Ronald William. *The Piltdown Men*. London: Gollancz, 1972.

Millar, Ronald William. *The Piltdown Men: [a Case of Archaeological Fraud]*. St. Albans: Paladin Frogmore, 1974.

Mitford, Jessica and Peter Y. Sussman. *Decca: The Letters of Jessica Mitford*. London: Weidenfeld & Nicolson, 2006.

Mitford, Nancy, Heywood Hill and John Saumarez Smith. *The Bookshop at 10 Curzon Street: Letters between Nancy Mitford and Heywood Hill 1952–73*. London: Frances Lincoln, 2004.

Mitford, Nancy and Charlotte Mosley. *The Letters of Nancy Mitford: Love from Nancy*. London: Hodder & Stoughton, 1993.

Mitford, Nancy, Evelyn Waugh and Charlotte Mosley. *The Letters of Nancy Mitford and Evelyn Waugh* Penguin Modern Classics. London: Penguin, 2010.

Muggeridge, Malcolm. *Chronicles of Wasted Time: The Infernal Grove*. Vol. 2. London: Fontana, 1975.

Müller-Hill, Benno. *Murderous Science: Elimination by Scientific Selection of Jews, Gypsies, and Others, Germany, 1933–1945*. Oxford: Oxford University Press, 1988.

Murray, Stuart. *Beyond Tithing*. Carlisle: Paternoster, 2000.

Nilus, Sergei Aleksandrovich and Victor Emile Marsden. *Protocols of the Meetings of the Learned Elders of Zion*. London: The Britons, 1931.

Noakes, Jeremy. 'Nazism and Eugenics: The Background to the Nazi Sterilization Law of 14 July 1933'. In *Ideas into Politics: Aspects of European History 1880–1950*, edited by Roger Bullen, H. Pogge von Strandmann and Antony Polonsky. London: Croom Helm, 1984.

Nottingham, Chris. *The Pursuit of Serenity: Havelock Ellis and the New Politics*. Amsterdam: Amsterdam University Press, 1999.

'Obituary: Lady Rumbold: Hermione Baddeley's Daughter and Society Beauty Who Enchanted the Young Lucian Freud'. *The Telegraph.* http://www.telegraph.co.uk/news/obituaries/3815125/Lady-Rumbold.html (accessed 9 October 2014).

Olby, Robert. 'Huxley, Sir Julian Sorell (1887–1975)'. *Oxford Dictionary of National Biography* (2004). http://www.oxforddnb.com/view/article/31271 (accessed 20 April 2010).

Overy, Richard. *The Morbid Age: Britain between the Wars.* London: Allen Lane, 2009.

Paluch, Andrzej K. *Malinowski.* Wyd. 2nd ed. Warszawa: Wiedza Powszechna, 1983.

Paul, Diane. 'Eugenics and the Left'. *Journal of the History of Ideas* 45, no. 4 (1984): 567–590.

Peel, Robert A. *Marie Stopes, Eugenics and the English Birth Control Movement: Proceedings of a Conference Organised by the Galton Institute, London, 1996.* London: Galton Institute, 1997.

Peterson, Houston. *Havelock Ellis, Philosopher of Love.* Boston, New York: Houghton Mifflin Company, 1928.

'Pitt-Rivers Collection for Museum'. *The Times,* 24 January 1975, 3.

Pitt-Rivers, George. 'The Sick Values of a Sick Age: A Response to Anthony Ludovici'. *The New Age* 26, no. 2 (1919): 25–27.

Pitt-Rivers, George. *The World Significance of the Russian Revolution.* Oxford: Basil Blackwell, 1920.

Pitt-Rivers, George. 'Correspondence'. *The Eugenics Review* XII, (1920–1921): 71–73.

Pitt-Rivers, George. 'A Psychological Study of the Artist and His Art'. In *The Art of George W. Lambert, A.R.A,* 31–40. Sydney: Art in Australia limited, 1924.

Pitt-Rivers, George. 'The Effect on Native Races of Contact with European Civilisation'. *Man* 27, (1927): 2–10.

Pitt-Rivers, George. 'Sex Ratios and Cultural Contact'. *Man* 27 (1927): 59–60.

Pitt-Rivers, George. 'Depopulation in Melanesia'. *Man* 28 (1928): 213–215.

Pitt-Rivers, George. 'Papuan Criminals and British Justice'. *Man* 29 (1929): 21–22.

Pitt-Rivers, George. 'Anthropological Approach to Ethnogenics: A New Perspective'. *Human Biology* 4, no. 2 (1932): 239–251.

Pitt-Rivers, George. 'Anthropological Approaches to Ethnogenics'. In *Essays Presented to C. G. Seligman,* edited by E. E. Evans-Pritchard, Raymond Firth,Bronislaw Malinowski and Isaac Schapera, 241–253. London: K. Paul Trench Trubner & Co. Ltd., 1934.

Pitt-Rivers, George. 'The Tithe Proposals'. *The Times,* 13 March 1936, 10.

Pitt-Rivers, George. 'The Truth About Spain: A Remarkable Report'. *The Anglo-German Review* 1, no. 4 (1937): 157–159.

Pitt-Rivers, George. *Your Home is Threatened!* Gillingham: T.H. Brickell & Son, 1939.

Pitt-Rivers, George. *The Story of the Ancient Manor of Hinton St Mary.* Farnham: The Pitt-Rivers Museum, 1947.

Pitt-Rivers, George Henry Lane Fox. 'Aua Island: Ethnographical and Sociological Features of a South Sea Pagan Society'. *The Journal of the Royal Anthropological Institute of Great Britain and Ireland* 55 (1925): 425–438.

Pitt-Rivers, George Henry Lane Fox. *The Clash of Culture and the Contact of Races: An Anthropological and Psychological Study of the Laws of Racial Adaptability, with Special Reference to the Depopulation of the Pacific and the Government of Subject Races*. London: George Routledge, 1927.

Pitt-Rivers, George Henry Lane Fox. *Weeds in the Garden of Marriage*. London: N. Douglas, 1931.

Pitt-Rivers, George Henry Lane Fox. *Problems of Population*. London: George Allen & Unwin Ltd., 1932.

Pitt-Rivers, George Henry Lane Fox. *The Czech Conspiracy: A Phase in the World-War Plot*. London: Boswell Pub. Co., 1938.

Pitt-Rivers, George Henry Lane Fox. *The Riddle of the 'Labarum' and the Origin of Christian Symbols*. London: Allen & Unwin, 1966.

Pitt-Rivers, George Henry Lane Fox. 'Editor's Preface'. In *Problems of Population*, edited by George Henry Lane Fox Pitt-Rivers, 7–9. Port Washington, NY: Kennikat Press, 1971.

Pitt-Rivers, Julian Alfred. *The Story of the Royal Dragoons, 1938–1945*. London: Published for the Royal Dragoons by W. Clowes, 1945.

Pitt-Rivers, Julian Alfred. *The People of the Sierra*. London, Beccles: William Clowes and Sons, Ltd., 1954.

Pitt-Rivers, Michael. *Dorset. A Shell Guide*. New ed., Edited by Paul Nash. London: Faber, 1966.

Pitt-Rivers, Michael. 'Cultural General'. *Books and Bookmen* 22, no. 9 (1977): 23–25.

Popenoe, Paul. 'Public Opinion on Sterilization in California'. *Eugenical News* XX, no. 5 (1935): 73, 75.

Potter, Marguerite. 'What Sealed Baldwin's Lips?' *Historian* 27, no. 1 (1964): 21–36.

Prucha, Václav. 'The Economy and the Rise and Fall of a Small Multinational State: Czechoslovakia, 1918–1992'. In *Nation, State and the Economy in History*, edited by Alice Teichova and Herbert Matis, 181–196. Cambridge: Cambridge University Press, 2003.

Pugh, Martin. *'Hurrah for the Blackshirts!': Fascists and Fascism in Britain between the Wars*. London: Jonathan Cape, 2005.

Race and Culture. London: Royal Anthropological Institute of Great Britain and Ireland, Institute of Sociology, 1936.

Ratner-Rosenhagen, Jennifer. *American Nietzsche: A History of an Icon and His Ideas*. Chicago, IL; London: University of Chicago Press, 2012.

Rehak, David. *Hitler's English Girlfriend: The Story of Unity Mitford*. Stroud: Amberley, 2011.

Reid, Brian Holden. *J.F.C. Fuller: Military Thinker*. Studies in Military and Strategic History. Basingstoke: Macmillan, 1987.

'Reply to the Council of E.E.S.' *The Eugenics Review* XII, no. 1 (1920–1921): 73.

Report of the Ninth Conference of the International Federation of Eugenic Organizations, Ninth Conference of the International Federation of Eugenic Organizations, edited by C.B.S. Hodson. Farnham: I.F.E.O., 1930.

Ribuffo, Leo P. *The Old Christian Right: The Protestant Far Right from the Great Depression to the Cold War*. Philadelphia: Temple University Press, 1983.

Rivière, Peter. *A History of Oxford Anthropology*. Methodology and History in Anthropology. New York: Berghahn Books, 2007.

Roberts, Noel. *From Piltdown Man to Point Omega: The Evolutionary Theory of Teilhard De Chardin*. Studies in European Thought. New York, Oxford: P. Lang, 2000.

Robinson, Paul A. *The Modernization of Sex: Havelock Ellis, Alfred Kinsey, William Masters, and Virginia Johnson*. Ithaca, NY: Cornell University Press, 1989.

Rose, June. *Marie Stopes and the Sexual Revolution*. Stroud: Tempus, 2007.

Rowbotham, Sheila and Jeffrey Weeks. *Socialism and the New Life: The Personal and Sexual Politics of Edward Carpenter and Havelock Ellis*. London: Pluto Press, 1977.

Russell, Miles. *The Piltdown Man Hoax: Case Closed*. Stroud: History, 2012.

Sawday, Jonathan. '"New Men, Strange Faces, Other Minds": Arthur Keith, Race and the Piltdown Affair (1912–53)'. In *Race, Science and Medicine, 1700–1960*, edited by Waltraud Ernst and Bernard Harris. London: Routledge, 1999.

Schaffer, Gavin. '"Like a Baby with a Box of Matches": British Scientists and the Concept of "Race" in the Inter-War Period'. *The British Journal for the History of Science* 38, no. 3 (2005): 307–324.

Schaffer, Gavin. '"'Scientific' Racism Again?": Reginald Gates, the Mankind Quarterly and the Question of "Race" in Science after the Second World War'. *Journal of American Studies* 41, no. 2 (2007): 253.

Schaffer, Gavin. *Racial Science and British Society, 1930–62*. Basingstoke: Palgrave Macmillan, 2008.

Searle, G. R. *Eugenics and Politics in Britain, 1900–1914*. Leyden: Noordhoff International Publishing, 1976.

Searle, G. R. 'Eugenics and Class'. In *Biology, Medicine and Society, 1840–1940*, edited by Charles Webster, 217–242. Cambridge: Cambridge University Press, 1981.

Seligman, C. G., F. R. Barton and E. L. Giblin. *The Melanesians of British New Guinea*. Cambridge: The University Press, 1910.

'Serious Riots in Johannesburg'. *Wanganui Chronicle*, 7 July 1913, 5.

Sharpe, Catherine. 'Hitler Girls in the New Germany'. *The Anglo-German Review* II, no. 1 (1937): 17–18.

Sharpe, Catherine. 'Shadow over Czechoslovakia'. *The Anglo-German Review* 1, no. 5 (1937): 202–203.

Simpson, A. W. Brian. *In the Highest Degree Odious: Detention without Trial in Wartime Britain*. Oxford: Clarendon Press, 1992.

Skidelsky, Robert Jacob Alexander. *Oswald Mosley*. London, New York: Macmillan, 1975.

Soloway, Richard A. *Demography and Degeneration: Eugenics and the Declining Birthrate in Twentieth-Century Britain*. Chapel Hill: University of North Carolina Press, 1990.

Soloway, Richard A. 'Blacker, Carlos Paton (1895–1975)'. *Oxford Dictionary of National Biography* (2004). http://www.oxforddnb.com/view/article/47726 (accessed 26 July 2010).

Spencer, Frank and British Museum (Natural History). *The Piltdown Papers 1908-1955: The Correspondence and Other Documents Relating to the Piltdown Forgery*. London: Oxford University Press: British Museum (Natural History), 1990.

Spencer, Frank and Ian Langham. *Piltdown: A Scientific Forgery*. London: Natural
   History Museum Publications, 1990.
Spurling, Hilary. *The Girl from the Fiction Department: A Portrait of Sonia Orwell*.
   London, New York: Hamish Hamilton, 2002.
Stern, Alexandra. *Eugenic Nation: Faults and Frontiers of Better Breeding in
   Modern America*. Berkeley: University of California Press, 2005.
Stevenson, Alice. '"We Seem to Be Working in the Same Line": A.H.L.F. Pitt-Rivers
   and W.M.F. Petrie'. *Bulletin of the History of Archaeology* 22, no. 1 (2012): 4–13.
Stocking, George W. *Malinowski, Rivers, Benedict and Others: Essays on Culture
   and Personality*. Madison: University of Wisconsin Press, 1986.
Stocking, George W. *After Tylor: British Social Anthropology, 1888–1951*.
   Madison: University of Wisconsin Press, 1995.
Stone, Dan. 'An "Entirely Tactless Nietzschean Jew": Oscar Levy's Critique of
   Western Civilization'. *Journal of Contemporary History* 36, no. 2 (2001):
   271–292.
Stone, Dan. 'The English Mistery, the BUF, and the Dilemmas of British Fascism'.
   The Journal of Modern History 75, no. 2 (2003): 336–358.
Stone, Dan. 'Nazism as Modern Magic: Bronislaw Malinowski's Political
   Anthropology'. *History and Anthropology* 14, (2003): 203–218.
Stopes, Marie Carmichael. *Married Love: A New Contribution to the Solution of
   Sex Difficulties*. 7th ed. London: G. P. Putnam's Sons, 1919.
Strobl, Gerwin. *The Germanic Isle: Nazi Perceptions of Britain*. Cambridge:
   Cambridge University Press, 2000.
Sweet, Matthew. *The West End Front: The Wartime Secrets of London's Grand
   Hotels*. London: Faber, 2012.
Tata, Jamshed R. 'Rivers, Rosalind Venetia Lane Fox Pitt- (1907–1990)',
   Oxford Dictionary of National Biography. http://www.oxforddnb.com/view/
   article/57570 (accessed 13 June 2014).
Tata, Jamshed R., 'Rosalind Venetia Pitt-Rivers', The Royal Society. http://thyroid.
   org/wp-content/uploads/2012/04/Rosalind_Venetia_Pitt-Rivers.pdf (accessed
   13 June 2014).
Taylor, A. J. P. *Beaverbrook*. London: Hamilton, 1972.
Taylor, Catherine. *If Courage Goes: My Twenty Years in South African Politics*.
   Johannesburg: Macmillan, 1976.
Thackeray, William Makepeace. *Vanity Fair*. London, 1848.
Thatcher, David S. *Nietzsche in England, 1890-1914: The Growth of a Reputation*.
   Toronto: University of Toronto Press, 1970.
'The Theatres'. *The Times*, 6 September 1928, 10.
Thompson, Laura. *Life in a Cold Climate: Nancy Mitford: A Portrait of a
   Contradictory Woman*. London: Review, 2003.
'Three Men Sent to Prison'. *The Times*, 25 March 1954, 4.
Thurlow, Richard C. *Fascism in Britain: A History, 1918–1985*. Oxford, UK,
   New York, NY: Blackwell, 1987.
Toynbee, J. M. C. 'A New Roman Mosaic Pavement Found in Dorset'. *The Journal
   of Roman Studies* 54, no. 1–2 (1964): 7–14.
Turda, Marius. *Modernism and Eugenics*. New York: Palgrave Macmillan, 2010.
Turda, Marius and Paul Weindling. 'Eugenics, Race and Nation in Central and
   Southeastern Europe, 1900-1940: A Historiographic Overview'. In *'Blood and*

Homeland': Eugenics and Racial Nationalism in Central and Southeast Europe, 1900–1940, edited by Marius Turda and Paul Weindling. Budapest: Central European University Press, 2007.

Vere, Francis. *The Piltdown Fantasy*. London: Cassell, 1955.

Visser, Wessel. 'The South African Labour Movement's Responses to Declarations of Martial Law, 1913–1922'. *Scientia Militaria: South African Journal of Military Studies* 31, no. 2 (2003): 142–157.

Waddington, G. T. '"An Idyllic and Unruffled Atmosphere of Complete Anglo-German Misunderstanding": Aspects of the Operations of the Dienststelle Ribbentrop in Great Britain, 1934–1938'. *History* 82, no. 265 (1997): 44–72.

Waterston, David. 'The Piltdown Mandible'. *Nature* 92, no. 2298 (1913): 319.

Wayne, Helena, ed. *The Story of a Marriage: The Letters of Bronislaw Malinowski and Elsie Masson*. Vol. 2. London: Routledge, 1995.

'Weeds in the Garden of Marriage'. *Nature* 129, no. 3257 (1932): 493.

Weiner, J. S. *The Piltdown Forgery*. London, New York: Oxford University Press, 1955.

Weingart, Peter. 'German Eugenics between Science and Politics'. *Osiris* 5, no. 1 (1989): 260.

Welshman, John. 'Eugenics and Public Health in Britain, 1900-40: Scenes from Provincial Life'. *Urban History* 24, no. 1 (1997): 57–75.

Werskey, Paul Gary. 'British Scientists and "Outsider" Politics, 1931–1945'. *Science Studies* 1, no. 1 (1971): 67–83.

West, Rebecca. *The Meaning of Treason*. London: Macmillan & Co. Ltd., 1949.

Wildeblood, Peter. *Against the Law*. London: Weidenfeld and Nicolson, 1956.

Wise, Leonard. *Mosley's Blackshirts: The Inside Story of the British Union of Fascists, 1932–1940*. London: Sanctuary, 1986.

Witkowski, Jan A. and John. R. Inglis, eds. *Davenport's Dream: 21st Century Reflections on Heredity and Eugenics*. Cold Spring Harbor: Cold Spring Harbor Laboratory Press, 2008.

Wrench, Evelyn. *Francis Yeats-Brown: 1886–1944*. London: Eyre & Spottiswoode, 1948.

Wright, Patrick. *The Village that Died for England: The Strange Story of Tyneham*. London: Vintage, 1996.

Young, Michael W. 'The Careless Collector: Malinowski and the Antiquarians'. In *Hunting the Gatherers: Ethnographic Collectors, Agents and Agency in Melanesia, 1870s-1930s*, edited by Michael O'Hanlon and Robert Louis Welsch, 181–202. New York, Oxford: Berghahn Books, 2000.

Young, Michael W. *Malinowski: Odyssey of an Anthropologist, 1884–1920*. New Haven, London: Yale University Press, 2004.

Ziegler, Philip. *King Edward VIII: The Official Biography*. London: Collins, 1990.

INDEX

www.ingramcontent.com/pod-product-compliance
Lightning Source LLC
Chambersburg PA
CBHW071855270326
41929CB00013B/2240